There's a story about farming in the world today.

It's not a good one.

Why not?

Because good stories are bad for headlines.

Bad stories get the attention.

They scare people into taking action.

When people are scared, they spend money.

When people sense risk, they'll warn others.

And we end up with sound bites, clickbait and widespread misinformation.

Farmers haven't been out there talking about what they do. They're busy feeding the world, and learning, growing and improving how they do it, safely and responsibly, for people and the planet.

It hit Dennis Bulani at an event with hundreds of business leaders from all over the world, when the conversation on stage turned to food, our food system and the farmers behind it. It felt like an attack on his ethics, his industry and the products they work hard to put onto our plates.

It was enough to make him speak up about the topics causing so many people concern. Enough to write this book about why you can trust the food on your plate.

WHAT A FARMER WANTS YOU TO KNOW ABOUT FOOD

What a FARMER wants you to know ABOUT FOOD

DENNIS BULANI

ULTIMATE
YIELD PRESS

Saskatoon, SK, Canada

ULTIMATE
YIELD PRESS

203 47th Street East, Saskatoon, SK, Canada • 306-477-1800

Hardcover: ISBN 978-1-0691042-0-5
Paperback: ISBN 978-1-0691042-1-2
Ebook: 978-1-0691042-02-9
Audiobook: ISBN 978-1-0691042-3-6

First Edition, 2024

Publisher's Cataloging-in-Publication Data

Names: Bulani, Dennis, author.
Title: What a farmer wants you to know about food / Dennis Bulani.
Description: Includes bibliographical references. | Saskatoon, Saskatchewan, Canada: Ultimate Yield Press, 2024.
Identifiers: ISBN: 978-1-0691042-0-5 (hardcover) | 978-1-0691042-1-2 (paperback) | 978-1-0691042-02-9 (ebook) | 978-1-0691042-3-6 (audio)
Subjects: LCSH Agriculture. | Food. | Food safety. | BISAC TECHNOLOGY & ENGINEERING / Food Science / Food Safety & Security | TECHNOLOGY & ENGINEERING / Agriculture / General | TECHNOLOGY & ENGINEERING / Food Science / Chemistry & Biotechnology | TECHNOLOGY & ENGINEERING / Food Science / Food Packaging & Processing
Classification: LCC S494.5 .B85 2024 | DDC 630--dc23

Cover and book design: Patricia Bacall, bacallcreative.com

Dedication

Grocery shoppers, this book is for you. For all the time you spend thinking, and maybe worrying, about how to choose the best food for your family that fits your budget. My hope for you is to find the fun in food shopping again; to walk the aisles with confidence knowing the care, craftsmanship and science behind the food we farm.

Farmers, this book is also for you. For committing your life, your family and your livelihood to feeding others, and doing so, often under the microscope of critics. I wrote this book to build a bridge from farm to plate, to tackle some of the big topics and make it easier to have balanced conversations about food. I hope it helps you tell the story of your farm and the food you produce.

CONTENTS

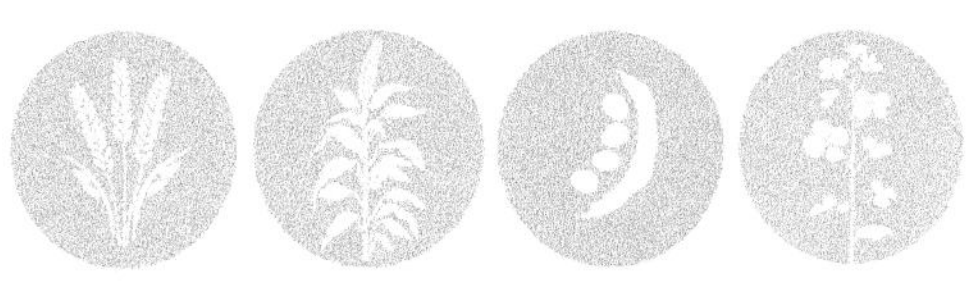

Chapter 1

WHY I WROTE THIS BOOK

It broke my heart. I was at an exclusive networking conference for entrepreneurs led by an individual for whom I have enormous respect. One of the guest speakers entered a dialogue with the conference leader about a supposed conspiracy between farmers, health insurance companies, pharmaceutical companies, and the media designed to keep people sick and dependent on costly medications to make ever-soaring profits. He claimed the food industry was deliberately making people obese, infertile, and depressed.

In response, the conference organizer told the audience that whenever he traveled, he brought eggs, olive oil, and other food items with him because he didn't think the food supply system was safe—in fact, he thought it was poisoning people as if we were still living in Upton Sinclair's jungle! And now, here he was, encouraging a roomful of highly educated and accomplished people, who trusted everything he said, to believe the same thing.

I paid a hefty price to attend this event where my life's work and my family's contributions to feeding and nourishing people for generations were denigrated and maligned. And the type of misinformation propagated in that presentation is becoming increasingly widespread and accepted.

I've been a farmer all my life. The average farmer in the US and Canada feeds 166 people. I may never have met many of the people who eat the food that comes from my farm, but my children and grandchildren eat those foods. Do you think farmers want to poison their loved ones—or anyone else—just to make a buck or because some C-suite executive for a pharmaceutical company told them to? Of course not. That's just absurd.

We all want to do the right thing—for our health, our bodies, and the planet. We want to provide nutritious, ethical, sustainable food for our families, and we want to do it with minimal harm to the environment. It can be extremely difficult to separate the wheat from the chaff in discussions around health, diet, and food production.

Some claims made by the guest speaker are founded on seeds of truth, which helps make them sound believable, but the reasoning behind them and the conclusions he extrapolated from those statistics were patently false. I won't debunk all his claims here because that would be a book in itself, but I will share a couple of examples.

For example, the number of people, especially women, experiencing mental health issues has increased over recent decades. The guest speaker attributes this to unhealthy food and eating habits when, in reality, there are multiple, generally benign causes that explain this rise, including destigmatization, better access to mental health services and diagnoses, and increased socio-economic pressures from things like the cost-of-living crisis, rising unemployment, and COVID-19 quarantine measures.

Of course, it stands to reason that food and food production play vital roles in our health, but it is one factor among many in the constellation of factors that contribute to our overall health and wellbeing, including genetics, lifestyle, policy interventions, health care access, and structural issues. What's more, the food available to us as consumers is safer and more accessible than ever before.

There is also a growing tendency to moralize food production and consumption. We label foods as "healthy" and "unhealthy," "good" and "bad," or "natural" and "unnatural," when those terms are context dependent. For example, if you buy butter or olive oil, these could easily be labelled as "unhealthy" foods because they are essentially just fats, but in the context of a balanced diet, the use of a single source of fat for cooking, especially in moderation, isn't going to have a significant impact on your overall health. In fact, fats can help our bodies better absorb and process certain nutrients!

We see this moralization, particularly in discussions of genetically modified organisms (GMOs) and organic foods. Most people believe that organic food is better than nonorganic and are likely have a negative impression of GMOs, but this largely boils down to a lack of understanding of the science behind these foods.

For example, if I offered you snack and said you can choose a banana or a mystery food containing glucose, fructose, glutamic acid, histidine, alanine, lauric acid, yellow-orange E101, ethyl hexanoate, and ethane gas, which would you pick? The banana, right? What if I told you they are both bananas?

Likewise, which form of food production do you think has the potential to feed more people, organic or nonorganic? And which has the lower carbon footprint? The answers will probably surprise you.

Mistrust and misinformation around how food is produced poses a danger to us all. We are talking about the food that fuels our everyday lives. Food that is grown to help stop hunger and famine in developing countries. Food that brings us together around the table for Sunday dinners and holiday celebrations.

Making people afraid of mainstream food production and the health care system is hugely profitable. Indeed, the guest speaker at that conference started a company that partners with supplement brands, fitness centers and gyms, telehealth services, so-called health food suppliers, and health and fitness technology—businesses worth millions of dollars. And he benefits from book sales, online subscriptions, and speaker fees on the same subject. He has a vested financial interest in driving consumer dollars toward his segment of the market just like other pundits in similar positions.

His goal to promote health was deeply entrenched and driven by a previous negative experience with a large company, which led him to assume all large companies were equally treacherous. For me, as a farmer with some degree of success, according to him that meant I was caught up in a nefarious conspiracy with Big Ag. His claims regarding our collective need for supplements, health aids, and particular lifestyle adaptations were based more on opinion, fearmongering, and demonization than fact.

I thought to myself, "*This is a smart guy with a goal to improve human health, but he has no idea what he's talking about when it comes to the intricacies of growing food. What he is missing is a farmer's perspective, someone who cares as much as he does about the safety of our food.*"

I felt insulted and sad that a fellow entrepreneur, as smart and successful as he was, hadn't taken the time to explore both sides of the story or considered asking a real expert and entrepreneur what was going on in the fields and factories that supply our food. It didn't sound like he had ever been to a modern farm or considered the host of people who dedicate their careers to researching these subjects, people like me who have come to a much different conclusion about it all than he had.

Remember, this was a networking conference for entrepreneurs! I love seeing entrepreneurs take a great idea and find success, but being successful in one area of expertise—say, public speaking—does not translate to being an expert in everything. There needs to be some degree of social responsibility that comes with success, especially for entrepreneurs with significant reach.

Moreover, the conversation can get pretty muddy pretty fast when it seems like everyone is out to turn a profit. Capitalism incentivizes demand. If I am an apple farmer, I am going to do everything I can to get you to buy *my* apples. Maybe that means pouring money into an expensive ad campaignor finding a way to make my apples sweeter, bigger, and juicier. It could even mean partnering with a big food wholesaler who can buy all my apples and sell them on to you. But does that mean I am going to secretly start adding things to my apples to make them addictive? The answer, clearly, is a resounding "No."

Part of the dilemma in the conversation around food production also rests on the fact that most people are more removed from the sources of the food they eat than ever before. For most of human history, people grew, raised, or hunted their own food, or they traded, sold, and bought at a relatively local level. Your grandparents or great-grandparents might have known their butcher, greengrocer, and fishmonger by name. Perhaps they even canned their own fruits and vegetables from their gardens.

That has all changed with the global market. Today, most people will never have the experience of growing their own food, aside from maybe a small herb garden in a kitchen window box. They won't know how quickly lettuce grows or the effect of heavy rain on certain crops. They won't know what goes into raising chickens or how their milk ends up in that carton in their fridge. And they won't know how heavily regulated every step of those processes is!

And that's okay. It is a good thing that farmers can provide so much food that it frees people to do all sorts of other important work that creates the fabric of our society and economy. In fact, less than two percent of the population are farmers in the US and Canada, which means that 98 percent of Americans don't have to spend their days toiling in the soil!

Still, I think many people continue to picture a farmer as a kind of country bumpkin, someone in overalls and a straw hat being manipulated by the whims of a faceless, evil mega-corporation that doesn't consider ethics, morals, or safety. Meanwhile, the industry has moved away from that stereotype by leaps and bounds. And this lack of knowledge and understanding creates doubt and mistrust.

Instead, it is more useful to think of farmers as specialists in their area of expertise. How often do people who are concerned about food ever ask a farmer? After all, we're the ones who grow it all.

We spend our lives, often generation after generation, learning everything about our industry, finding ways to improve it, and producing data-based solutions to real-life problems.

Would you rather have a top heart surgeon conduct your bypass surgery or someone who has seen every episode of *Grey's Anatomy*? There is a reason we trust experts with big decisions and difficult tasks, and I believe there are more good, honest people—in society at large and every industry within it—who care about their friends and families and extend that care and concern to others. I know that personally, I care deeply about these issues, and that should give you a sense of security when it comes to food production.

There are real challenges ahead with food production. The population is projected to hit nine billion people by 2050, which means we will have another billion people to feed in the next 35 years. In turn, food production will need to increase by 35 to 56 percent.[1] That is a monumental task if we want to maintain global food security, and it's not one that can be solved by small organic farms or bureaucratic red tape that creates barriers out of fear.

I think it is time to rebuild the relationship and understanding between producer and consumer. I want to fill in those knowledge gaps. I want people to understand where their food comes from, how it is produced, and everything that happens in that journey from farm to table. Food production is a complex system, but one I hope to make more accessible and less opaque to you with this book.

I've worked in the agriculture industry for over 40 years. My great-grandfather emigrated from what was then Russia (now Ukraine), and in 1914, the Canadian government offered a scheme whereby if you broke 22 acres of land, that land would become yours. So, he plowed the soil and planted wheat because coming from the "old world," that was all he knew how to grow.

That same farm passed down through four generations of my family until I took it over in 1981. I earned a Bachelor of Science degree where I studied animal nutrition, soil science, and ag economics. I've served as a keynote speaker at conferences nationally and internationally.

And my wife, Lynda, and I founded The Rack, which started out by offering fertilizer supply and crop protection services and grew into an industry leader in the science of agriculture and agronomic knowledge.

---

1 van Dijk, M., Morley, T., Rau, M.L. *et al.* A meta-analysis of projected global food demand and population at risk of hunger for the period 2010–2050. *Nat Food* 2, 494–501 (2021). https://doi.org/10.1038/s43016-021-00322-9

I have three daughters who are likely to follow in our footsteps and take over leadership of the farm, perhaps even passing it to their children, the sixth generation.

I feel immense pride that I'm playing a role that helps feed the world and that what I do will impact future generations. There is a special kind of satisfaction in knowing that when I harvest a crop, it will end up on someone's table. Over the years, I have grown wheat, rye, barley, canola, peas, lentils, flax, fava beans, and pinto beans, crops that fuel and nourish people. And I want to leave the world a better place than I found it.

I've written this book to demonstrate that so much of what people take for granted about food, including farming, organic labels, GMOs, pesticides, preservatives, and other vital subjects is just plain wrong. There is so much misinformation out there. I want to equip you with the facts so you can judge for yourself. Our food supply system has never been safer, and never have farmers been able to feed so many people.

I have spent my lifetime asking all the same questions you ask. This book is the result of my research and conclusions from that research. I want to share with you, in a condensed fashion, the same the road I have taken to better understand food production. I have tried to be as objective as possible and base my conclusions on my experience, the science I use to farm, and peer-reviewed research conducted by experts across disciplines and published in reputable sources.

They say the truth will set you free. I wrote this book so you can learn the truth about food. I've written it in the form of vital questions, some of which may have crossed your mind from time to time. Read the ones that matter most to you, or better still, read them all! When you get to the end of the book, I hope you'll agree that our food system, despite what that conference leader might believe, is safe and that, by and large, you can trust what comes out of the ground. At a bare minimum, you will understand why I believe this.

Chapter 2

CAN FOOD BE TRUSTED?

It has crossed my mind that more people might possibly die from the stress of determining what they should eat than from the food they eat. Of all the things that we have to stress about in our lives, the food on our shelves does not need to be one.

Let's get right into it—yes, our food production system is trustworthy. We use chemicals and herbicides with tactical precision. Food is safer than ever. Tighter regulations, new technologies, industry adoption of robust safety systems, and heightened public scrutiny mean that the global food supply chain is healthier than ever before.

However, mistrust of the food supply persists, perhaps because consumers don't clearly understand how we grow food and the processes and technologies involved. In this chapter, I'll walk you through the scientific evidence that distinctly demonstrates the safety of our food supply. You don't have to bring eggs and vegetables when you travel!

We farmers are doing everything we can to create a world with no starvation and maximum longevity and health, but before we can appreciate how safe our food is today, we need to look back at what made it unsafe before. What is at stake if we roll back the clock on food production and safety?

In the old days, families would preserve fruits, vegetables, and even meats as a long-term food supply during the winter months when crops could not grow. When my grandmother preserved tomatoes, she would boil the Mason jars in water to seal the jars and kill the bacteria. Why? Because not doing so could risk exposing anyone who ate her tomatoes that winter to botulism. Sometimes, whole families died from this deadly foodborne illness.[2]

Typhoid Mary[3] is perhaps the most famous example of how deadly foodborne illnesses were before the industry became as highly regulated as it is today. In the early 1900s, Mary Mallon worked as a cook for several families in New York City until it was discovered that she was a carrier of the bacteria *Salmonella typhi*. She is thought to have infected over 100 people with typhoid fever simply through casual exposure and poor sanitation.

Diving even further back in history we find a lesser known but even more impactful example of the need for food safety—the curious case of ergot poisoning and the French Revolution in 1789. Ergot fungus grows on rye and other grains, and when ingested by humans, it causes ergotism, with symptoms including mania, hallucinations, and psychosis. Indeed, the hallucinogenic drug lysergic acid diethylamide, better known as LSD, is related and similar in chemical structure to ergot. Historian Mary K. Matossian found evidence that severe outbreaks of ergotism played a role in "The Great Fear," a period of unrest and general panic in France that gave rise to riots, violence, and ultimately, the revolution that toppled the country's monarchy.

These may sound like extreme examples, but it is critical to understand just what is at risk when we talk about food safety. Much of the distrust and panic over food production comes from the good-natured but ill-guided belief that everything "natural" is better for you.

For example, organic food is perceived to be better because it seems more "natural," but botulinum toxins, salmonella, and E. coli are all-natural, too, and we take rigorous and deliberate measures to keep them

<hr>

2 "Family Botulism Deaths." *Prairie Public.* 28 June 2021. https://news.prairiepublic.org/dakota-datebook/2021-06-28/family-botulism-deaths

3 Marineli F, et al. "Mary Mallon (1869-1938) and the history of typhoid fever." *Annals of Gastroenterology.* 2013;26(2):132-134. https://www.ncbi.nlm.nih.gov/pmc/journals/2375/ See also: Connolly, Kevin. "How Typhoid Mary left a trail of scandal and death." *BBC.* 20 April 2020. https://www.bbc.co.uk/news/world-europe-52291327

from contaminating our food so people don't get sick. As food production became increasingly outsourced outside the home and more regulated, we developed modern technology to prevent foodborne illnesses like botulism, salmonella, and ergot from entering the food supply. We developed strict sterilization standards and preservatives to maintain that sterilized state.

Today, when you buy a can of tomatoes from the grocery store, you can be assured that there is some form of modern food preservative inside to prevent botulinum from growing in the container. Processes like pasteurization, irradiation, and better packaging mean reduced contamination risks, longer shelf life, and less food waste.

Still, many people demonize the very food preservatives that help us stay safe and healthy and, instead, would have us believe they are "unnatural" and, therefore, "unhealthy." The alternative, of course, is that we abandon preservatives and all risk getting botulism!

When we grow and process crops, we are constantly fighting Mother Nature—whether in the form of bacteria, fungi, or pests that feed off those same food sources. These organisms can destroy crops and harm us if ingested. We are in an endless battle to prevent crop loss and disease, and that has been true for all of human history. Written records dating back to the Assyrian Empire in 600 BCE reference the fight against crop disease and foodborne illness.[4]

For example, fusarium head blight is a fungus that grows in the head of cereal grains like wheat, oats, barley, and rye. It is deadly poisonous and devastating to the crop. So, when I grow wheat in my fields, if the weather is moist and humid when the wheat is flowering, it is likely that fusarium fungus will infect the wheat. And if my wheat has fusarium in it, I literally have to dig a hole and bury it. I lose the crop and any profits from it, but our food system remains safe.

Instead of risking the continued profitability of my business, the downstream consumer's health, and the security of our food production

---

4 Bennett, J.W.; Bentley, Ronald. "Pride and Prejudice: The Story of Ergot." *Perspectives in Biology and Medicine.* 1999. 42(3): 333-355. Doi: 10.1353/pbm/1999/0026

we treat crops with a protective fungicide that stops the fusarium head blight from growing in the first place.

More importantly, we only spray that fungicide at the time of year when the wheat is flowering and only if conditions are conducive to the formation of fusarium head blight. There are strict rules about when, how, and how much a chemical can be applied in commercial farming.

Think of it like taking Tylenol for a headache. You only take the recommended dose as advised in the instructions when you already have a headache. You wouldn't take the whole box just because you think you might get a headache next week.

Before I sell my wheat, I must sign a lengthy declaration testifying that I did not use specific chemical compounds in its production. This helps prevent cross-contamination when my wheat is shipped off with other farmers' crops to be processed. If I fail to take the required precautions, I am liable for any cross-contamination that occurs in transport. A single kernel of contaminated seed can bring down the whole truckload. That means tens of thousands of dollars and months of labor are lost. It is not worth the risk to farmers to take safety standards lightly.

In the now-rare cases when there is a product recall, commercially produced foods have RFID tags and bar codes that make them quickly identifiable and traceable. This allows us to remove contaminated foods from the supply chain faster.

GMOs or genetically engineered (GE) foods are another way that modern farmers protect crops against common pests and diseases. Some plants, like corn and cotton, have been genetically modified to repel insects by releasing pheromones. Others have essentially been inoculated against specific diseases, much like we are. This is why potato crops can now fend off the blight that led to the Great Famine in Ireland in the mid-1800s.

The idea behind this is that the plant protects itself instead of relying on the farmer, and farmers can tend to more acres with less equipment and fewer chemicals.

Part of the battle against foodborne illnesses is also a battle against decay. When you choose produce at a grocery store, it is unlikely you'll pick up a head of lettuce that's turning brown at the edges. This signifies it's beginning to rot—a process sped along by bacteria and fungi, which help break the lettuce down. Preservatives slow or even stop the breakdown of food and prevent it from rotting.

This is important from health and safety and logistics perspectives. Rot generally means harmful microorganisms are hard at work, and food now has further to travel between producer and consumer; 1,500 miles on average.[5] So, the longer we can keep food fresh and healthy, the more widely available nutritious food is to people in remote locations or places far from where food is grown.

And for consumers concerned about these "food miles" and the impact transporting food has on emissions and their carbon footprint, remember that *what* you eat has a greater impact on your carbon footprint than *where* it comes from.

Recent events like the COVID-19 pandemic and war in Ukraine have shown us just how important reliable, sustainable food production is, especially of staple crops like wheat, rice, and corn. Without that market stability and predictability, we all suffer. Farmers lose their income, consumers see the cost of staple goods rise, and people are forced to go hungry.

Everything we think we know about health and food safety depends on context. For example, organic farmers have used naturally occurring chemicals like spinosyn to treat their crops. Bacteria in the soil produces spinosyn and, therefore, it is listed as an organic insecticide. You can buy it at greenhouses, grocery stores, and even online through Amazon. However, it's known to kill bees for at least three hours after application, and there have been toxic poisoning events in humans related to spinosad

---

5 "Eating up the miles? Navigating the twists & turns of food transport." *Epigram.* The University of Bristol. 1 March 2018.

(spinosyn A and spinosyn D) exposure,[6] especially when combined with flonicamide,[7] another organic insecticide.

You won't find spinosad being used very often in commercial production because there are safer, more effective alternatives. On my farm, we sometimes spray Coragen, which kills some insect types but is much safer on all pollinators, including bees. In fact, after we spray, my neighbor—a commercial beekeeper—asks to set his hives up alongside my fields because canola flowers make great honey!

So, does using compounds like spinosad and flonicamide mean organic food is toxic? Of course not, but these compounds come with a nearly identical warning label to the non-organic glyphosate I use on my crops. This is because the question of safety comes down to toxicity, i.e., how much of a chemical or substance a person would need to ingest to experience the toxin's harmful effects. Both organic and non-organic farmers are subject to strict guidelines and regular testing to meet safety protocols in this regard.

Likewise, remember the conference host from Chapter 1 who carried olive oil with him out of fear of ingesting toxins? He had fallen victim to myths and misinformation about the risks relating to cooking oils like canola. So, what is the real story here? Well, it is true that pretty much anything you burn at a high enough temperature creates free radicals, including something as benign as canola oil, but that can easily be prevented by choosing the right cooking oil for your dish and heating it to the right temperature. Contrary to some claims, the temperatures used to extract canola oil is carefully monitored to stay low enough for the oil to be extracted without creating toxins from high heat.

Oils typically used for deep frying, like canola, corn, and even virgin olive oil, have very high smoke points—upwards of 400 degrees. And if

6 Hamad MH, et al. "Acute poisoning in a child following topical treatment of head lice (pediculosis capitis) with an organophosphate pesticide." *Sudanese Journal of Paediatrics.* 2016;16(1):63-6. PMID: 27651556; PMCID: PMC5025936.

7 Su T-Y, et al. "Human poisoning with spinosad and flonicamid insecticides." *Human & Experimental Toxicology.* 2011;30(11):1878-1881. doi:10.1177/0960327111401639 COPY CITATION

you've ever left a pan burning on the stove to the point where something starts smoking, you'll know it for sure! It's a mistake you're not likely to repeat. Abandoning canola oil—or any commercially produced oil—for this reason is a bit like setting off fireworks facing your house and being surprised when it catches on fire. If you follow the instructions for use, there is no danger.

In other words, fears drummed up about cooking oils and toxins are generally unfounded. As we'll see, that is the case with a lot of the fears people have about food safety.

We'll do a deep dive on the individual subjects of GMOs, organic farming, and use of fertilizers and preservatives, but first, I want to briefly address here why not all food can be GMO-free, organic, and pesticide- and preservative-free because that is a big part of the question "Is our food safe?" for a lot of people.

The short answer is simple: We could not possibly feed everyone on the planet if that were the case. Organic food production is inefficient. It has much lower yields than farming and production methods that employ modern technology, and thankfully, we do not need our food to be all those things to be safe. The science is clear on this. We are living longer, healthier lives.

For example, GMO foods are more widely consumed in North America than in Europe and have been for decades, and yet, researchers have found no evidence of GMO-linked increases in cancer, obesity, autism, or other key health indicators as they compare with our European counterparts.[8]

In fact, the rates of new cases of all cancers and cancer-related deaths have been steadily declining for the last three decades, according to the National Institutes of Health.[9] To put it plainly, no one has ever died or grown a third foot from eating a GMO food.

---

8 Ducharme, Jamie. "What the science says about gmo foods." *Time Magazine*. 13 January 2024. https://gb.readly.com/magazines/time-magazine-europe/2024-01-13/65a03aca05f117cfd909fe08
9 "Cancer Stat Facts: Cancer of Any Site." National Cancer Institute. National Institutes of Health. https://seer.cancer.gov/statfacts/html/all.html

The perception of organic farming is also shrouded in its own mythology. Consumers hear "organic farming," and they picture a bucolic scene of a small, independent farmer, hand-picking strawberries. The reality is that most organic food is mass-produced nowadays and sold to multinational corporations like Costco and Whole Foods. It is a big business producing supposedly healthier food, which can be sold at four times the price.

The broad reason most commercial food producers use the technology available to us—including GMO plants, fungicides, pesticides, fertilizers, and preservatives—is that they are proven to be safe and are absolutely essential for feeding an ever-growing population. It also means that more of us can enjoy a healthy, balanced diet that includes a variety of foods.

For most of human history, the range of fruits, vegetables, whole grains, meats, and other products we eat daily would have been entirely unavailable, extremely rare, or accessible only to the wealthy elite. A single pineapple cost the equivalent of $8,000 in today's money just 200 years ago! I don't know about you, but I don't want to go back to a time when only kings and queens could afford to enjoy things like coffee, meat, and spices.

It comes down to the simple economics of supply and demand. We use modern technology so that we can produce more food, more safely, in a predictable, sustainable way, and that lowers the cost to consumers by increasing the supply.

At the heart of modern farming is the ability to fight the aspects of Mother Nature that threaten our food security and amplify those that benefit it. What do I mean by this? Well, we've talked about the kinds of things that can be harmful to humans if left unchecked, but what about natural processes that help grow our food? Modern technology has given us the power to mimic and, in some cases, augment those processes.

For example, the number one ingredient in most fertilizers is nitrogen. It is one of the most abundant elements in the universe, and although it makes up about 78 percent of our atmosphere, it is exceedingly rare in solid parts of the Earth. That is bad news for plants because plants need nitrogen for photosynthesis.

Plants love nitrogen so much that they basically use up every molecule in the soil during one season. In the old days, farmers would rotate crops, also called "summer fallow," or spread manure on their fields. Crop rotation with summer fallow (cultivation) means losing one year's worth of that crop from a given field, and manure and other "organic" fertilizers do add some nitrogen and micronutrients back into the soil, but they do so at an extremely inefficient rate such that the soil eventually degrades.

In contrast, modern commercial farmers use fertilizers to enrich the soil we grow crops in. Nitrogen is the No. 1 ingredient in those fertilizers because it is considered the single most important element for plant growth, development, and yield.

This last part—yield—is essential. Nitrogen as a fertilizer is credited with a significant increase in food production globally, and one that has saved countless people from hunger, malnutrition, and death.[10] In commercial agriculture, if we use nitrogen-rich fertilizer, we can produce three to five times the yield we would get without it. That's a lot of mouths we're feeding, and at the same time, we are replenishing the soil system, which would otherwise be mined.

Finally, I think many people don't realize how much testing goes on in food production and processes. Millions of dollars are spent on research and testing before fertilizers and crop protection products even hit the market to ensure they are safe. Academics and experts around the world are dedicated to researching and testing the safety of our food and new agricultural technology. We have to prove that any chemicals and additives used at any stage of food production are safe for consumption and at specific dose rates.

We also test the soil crops are grown in. We test foods for chemical composition, pathogens, and the potential presence of unsafe ingredients. Our farms and premises are subject to stringent regulations and inspections

---

10 Prasad, R., Shivay, Y.S. A Brief History of the Fertilizer Nitrogen. *Indian Journal of History of Science.* 56, 60–64 (2021). https://doi.org/10.1007/s43539-021-00006-0

for hygiene and food standards. We submit foods to further sampling by buyers, regulators, and third-party food safety authorities.

If we apply Occam's Razor to the question of whether our food supply is safe, the answer is a simple but resounding "Yes." There is no diabolical conspiracy to make people sick or control you from the inside out by genetically modifying your DNA. Farmers want to do their job—we want to feed people. We want to keep people from starvation and malnutrition, we want to leave the Earth better than we found it and, yes, we want to make our lives easier, too. Would you rather pick all the crumbs out of your carpet by hand or use a vacuum? Agricultural technology is no different.

It comes down to economies of scale. How can we feed and nourish the greatest number of people with the least impact on the environment and in the safest way possible? We use technology to produce safe, sustainable, predictable yields that feed seven billion humans and still counting! And we have a whole network of people and institutions holding us accountable. Most food is grown by family farms; some family farms are growing very large due to economy of scale.

Many people say big corporate farms are controlling our food. Well, the reality is that many large family farms have had to grow to be economical and remain in business. Is that bad? Why can any other business grow and thrive but when a farmer succeeds, it's considered unethical? Consumers seem to think farmers should be out on the back 40 in coveralls and a pitchfork because that is the perceived order. However, farmers today are professional businessmen and women, and the truly adept and skillful ones are getting bigger and more efficient. That certainly does not mean their family values and concern for producing safe food has changed.

Chapter 3

WHAT'S THE TRUTH ABOUT GMO?

What do you picture when you think of the first genetically modified food? Do you see a scientist in a white lab coat working for a monolithic, nefarious food conglomerate? What about early humans sowing seeds from a plant that produced the most kernels? The truth about genetically modified organisms, or GMOs, is that humans have been using selective breeding to produce desirable genetic adaptations in plants and animals for thousands of years! The only difference is that now, we can reliably and safely achieve the same outcome through gene editing techniques.

Before we dive into the myriad benefits of GMO foods—from enhanced nutrition to disease and drought resistance—we need to understand how this technology works and the long history of selective breeding it is built upon.

Scientists have found evidence of selective breeding and crossbreeding by humans as early as 8000 BCE. People planted the seeds from crops that produced the highest yield or tastiest fruit and bred animals that had the healthiest offspring or calmest tempers. Indeed, most food we eat today was created through such interventions in the so-called "natural order." It is how we got orange carrots and cuddly Labradoodles.

This process was largely trial and error for most of human history, and it took generations for a plant to predictably produce the desired traits. This is because although the farmer chose which parent plants to crossbreed or seeds to replant, the results were unpredictable at the genetic level because DNA recombines randomly. This element of randomness also meant that undesirable traits like lower yield or poor taste often resulted alongside the desirable traits like pest resistance.

Our collective understanding of genetic inheritance took a big leap forward in the early 20th century with the application of Gregor Mendel's research on pea plants. His findings served as the foundation for other scientists to use mathematical probabilities to predict the expression of traits in crossbred plants. You may remember filling out Punnett squares at school to show how genes are inherited and expressed.

Then, the discovery of the structure of DNA came in 1952, which set the stage for the next quantum leap in how we produce food. Just over 20 years later, in 1973, Herbert Boyer and Stanley Cohen inserted DNA from one bacterium into another, and thus, we had the first genetically engineered organism—though I can't imagine this one was particularly tasty.

It might surprise you to learn that the first GMO product available to consumers was insulin. That's right—a lifesaving drug was the first application of modern genetic engineering. Before that, from about the 1920s onward, people had to use insulin obtained from cow and pig pancreas', which is actually slightly different from human insulin. And before the 1920s, diabetes was essentially a death sentence.

But thankfully, modern genetic engineering changed that. Today, more than seven million Americans inject genetically engineered insulin daily[11] without a second thought, and that number skyrockets to between 400 and 500 million people globally.[12] In the over 40 years since it was introduced, not one person has grown a second head or a third eye because of it.

It was clear from the start that technological advances in genetic engineering were going to be game changers across industries, therefore, the

____

11 Locklear, Mallory. "Insulin is an extreme financial burden for over 14% of Americans who use it." *Yales News.* 5 July 2022. https://news.yale.edu/2022/07/05/insulin-extreme-financial-burden-over-14-americans-who-use-it

12 Basu, Sanjay, et al. "Estimation of global insulin use for type-2 diabetes, 2018-30: a microsimulation analysis." *The Lancet: Diabetes & Endocrinology.* January 2019. 7 (1): 25-55. doi: https://doi.org/10.1016/S2213-8587(18)30303-6

federal government and agencies like the Food and Drug Administration (FDA), the Department of Agriculture (USDA), and the Environmental Protection Agency (EPA) got to work on setting policy and regulating safety standards for future GMOs.

It took nearly a decade for the first GMO food, a tomato, to get through all the hoops of testing and evaluation to be deemed safe for consumers. That opened the floodgates, and a wave of GMO foods, including corn, potatoes, soybeans, and canola, were approved.

So, how are GMO foods made today? First, scientists identify which gene gives an organism—be it a plant, animal, or other microorganism—a particular desired trait, like higher yield or pest resistance. Once scientists identify this genetic information, they can copy it and insert it into the DNA of another organism. Advances in genetic technologies like CRISPR, which functions like gene editing scissors, make the process of silencing or activating targeted genes even more accessible.[13]

Then, scientists grow, monitor, and test the new genetically modified organism. It takes years of rigorous reviewing and testing a GMO before it enters the marketplace for public consumption.

For example, when scientists wanted to grow pest-resistant corn to reduce the need for pesticides to fight corn borers, they first identified a gene in a soil bacterium called *Bacillus thuringiensis* (Bt), which produces a natural insecticide.[14] Interestingly, the spores of this same bacterium are often sprayed as a pesticide in organic farming.

The next step in the process was for scientists to copy the gene responsible for producing the insecticide and insert it into the DNA of a corn plant. That's a change to one single gene out of roughly 32,000 genes in a corn plant. Then, scientists grew the new plant in a lab, monitored and

---

13 "What is CRISPR and How does it work?" *Livescience Tech.* 30 April 2018. https://www.livescience.tech/2018/04/30/what-is-crispr-how-does-it-work-is-it-gene-editing/

14 "Science and History of GMOs and Other Food Modification Processes." U.S. Food and Drug Administration. 3 May 2024. https://www.fda.gov/food/agricultural-biotechnology/science-and-history-gmos-and-other-food-modification-processes

field-tested it, and subjected it to in-depth reviews for health, safety, and pest resistance. Today, that plant is available as "Bt corn."

When I faced the choice of whether to grow GMO crops, I felt obligated to question my options before deciding. I perhaps took the question more seriously than an average consumer because I wanted to be absolutely certain that the food I grew for people would nourish and not harm anyone. Like every farmer I know, I feel a tremendous responsibility toward consumers, not least because my friends and family are among them.

I was considering planting GMO canola and wanted to know precisely where it came from. What would be the repercussions of planting it? Would it change the nutrition or safety of my crops? Would it affect my soil's health? How would it affect crop rotation? Would it have a negative or positive impact on my commitment to regenerative agriculture? I did my research. I asked all the same questions as consumers do and then some.

As I worked through my list of questions, it became clear that there were no valid reasons against planting a GMO crop. In fact, the positive aspects of such a change were overwhelming.

First and foremost, I am committed to having the healthiest possible soil because that is what we are leaving to future generations. That is why I practice regenerative agriculture and have done for nearly 40 years. As such, I follow certain rules regarding soil management, including zero tillage and direct seeding, with the goal of increasing the level of organic matter in the soil.

The kind of GMO I was considering was called "Roundup Ready" or glyphosate-tolerant canola because the plants were resistant to glyphosate herbicides, including the brand Roundup. If I planted Roundup Ready canola, I could spray Roundup on weeds to kill them without harming my crops instead of cultivating the weeds out, which would involve tilling and, therefore, damage the integrity of the soil.

That was a major upside for the health of the crops and the soil. It also translated to financial savings. I would not need to buy expensive herbicides and I would avoid the high cost of tillage, which burns a lot of fuel and requires a ton of horsepower to work the soil. In other words, I could lower the cost of growing food on my farm while improving the soil quality. Planting GMO canola would make my farm more sustainable. I decided it was the right choice.

But the big question that worries consumers is the health and safety of GMO foods for consumption. Perhaps it sounds too good to be true, but GMO foods are just that good. The science is very clear on this. GMO foods are scrupulously studied before they are sold to the public. There is no link between GMOs and genetic mutations in humans, GMOs do not affect fertility, pregnancy, or offspring, and there is no evidence of gene transfer between GMO foods and consumers.[15]

GMO foods are as safe as any other food we eat, and GMOs do not affect consumer health any differently than non-GMO foods—except for GMO foods that have been modified to increase their nutritional content, which are, therefore, healthier.[16]

Indeed, increased nutritional value is one of many benefits of GMO foods. Some other advantages include tastier food, plants resistant to disease, drought, and pests, reduced water, land, and fertilizer use for crops, less use of pesticides, increased supply of food, longer shelf life, and lower costs to produce food.[17]

A major upside is that GMOs also protect our environment. On my farm, soil erosion has decreased by 93% because we no longer need to till the soil to kill weeds; we can spray targeted herbicides without harming our crops. We have reduced herbicide and pesticide runoff into groundwater by 70% because we use only one type of herbicide at a very low rate, and GMO crops have allowed us to eliminate the use

15 Norris, Megan L. "Will GMOs Hurt My Body? The Public's Concerns and How Scientists Have Addressed Them." Harvard University. 10 August 2015. https://sitn.hms.harvard.edu/flash/2015/will-gmos-hurt-my-body/
16 "GMOs and Your Health." U.S. Food and Drug Administration. July 2022. https://www.fda.gov/media/135280/downloads
17 "Genetically modified organisms – GMOs." MedLine Plus, National Library of Medicine. 30 July 2022. https://www.medlineplus.gov/ency/article/002432.htm

of certain pesticides in the environment thanks to the simple manipulation of a gene.

Furthermore, greenhouse gas emissions have been reduced by 326 million pounds a year since the mid-1990s in the US alone. That is because tilling soil releases carbon into the atmosphere. When you direct seed or practice regenerative agriculture like I do, you don't disturb the soil, and the soil sequesters carbon. It is called a "carbon sink." On average, one acre of zero-tilled soil sequesters about one ton of carbon. GMO crops let us do that—we improve the environment when we use GMOs!

Yet there continues to be fear among consumers of some unknown or unforeseen consequence of eating GMO foods, and myths about GMO food safety persist in popular culture. Among the most common myths is that GMO foods cause health problems like allergies, celiac disease, and even cancer—none of which are remotely true.

We know these myths aren't true because GMOs are so carefully studied at every stage of development, and their effects continue to be studied long after they've entered the marketplace. For example, to create GMOs, scientists must first know which new proteins the plant will produce.[18] Then, they conduct thorough tests to ensure these proteins are not allergens or toxins. Also, FDA regulations ensure that foods derived from GMOs must be as safe for consumption as non-GMOs.

Some people have posited that cases of celiac disease are rising because of increased GMO consumption, but that isn't true. Celiac disease is caused by an abnormal immune response to gluten, which is a protein found in wheat, rye, and barley. No GMO wheat, rye, or barley is available to North American consumers, so there can be no correlation between GMO foods and celiac disease.

Lastly, scientists have disproved any link between GMO foods and cancer with countless peer-reviewed studies. Most recently, this has been

18 "GMOs and Your Health." U.S. Food and Drug Administration. July 2022. https://www.fda.gov/media/135280/download

proved by studying cancer rates in the US and Europe.[19] Americans eat more GMO foods than Europeans, so if there was a correlation between GMO foods and cancer, we would expect to see higher rates of cancer in the US and Canada when compared with Europe, and there is no such difference.

Such fears and myths generally stem from either a limited understanding of the science behind genetic engineering or the deliberate perpetuation of misinformation by people who stand to benefit from the genuine, albeit misguided, fears of others.

The truth is that genetic engineering is helping us in the fight against diseases. Genetic engineering has led to the development of antibiotic drugs, vaccines, and hormones that are used to prevent and treat a wide range of health issues, including cancer, heart disease, arthritis, hemophilia, hepatitis, stroke, and seizures.[20] And more disease-fighting applications are being developed every day.

Another reason some consumers are suspicious of GMO food producers is because they believe we are growing GMO foods solely for profit. While it is true that retailers, including myself, make money selling GMO seeds, that would be true regardless of whether those seeds were GMO or not. Farmers are in the business of selling crops to consumers. We have to make money if we want to stay in business and keep feeding people. It may sound absurd to point that out, but it is surprisingly easy for consumers to overlook.

Even so, growing GMO crops lowers the cost to farmers, as we've discussed. Those savings get passed on to consumers because we operate

19 "Genetically Engineered Crops: Experiences and Prospects- New Report." National Academies of Science, Engineering, and Medicine. 17 May 2016. https://www.nationalacademies.org/news/2016/05/genetically-engineered-crops-experiences-and-prospects-new-report

20 Tian, Xiuchun. "Pharmaceutical Use of GMOs." University of Connecticut, College of Agriculture, Health and Natural Resources. 9 November 2017. https://gmo.uconn.edu/topics/pharmaceutical-use-of-gmos. See also "Whole Genome Methods and Pharmaceutical Applications of Genetic Engineering." *Allied Health Microbiology*. Oregon State University. 2019. https://open.oregonstate.education/microbiology/chapter/12-1whole-genome-methods-and-pharmaceutical-applications-of-genetic-engineering

in a competitive economy. The fact that we can produce higher yields also means there is a greater supply, which further lowers the costs to consumers. It's a win-win.

Fearmongers also used to argue that GMO products would create monopolies, and farmers would be beholden to a small number of major corporations like Monsanto, DuPont, and Dow Chemical that held the patents. That claim quickly fell apart because patents on GMO products do not last very long. For example, Monsanto's patent on glyphosate ended in 2000, and now, there are a dozen or so glyphosate manufacturers worldwide. So, once a patent comes off, anyone can make, sell, and profit from these technologies.

Incidentally, retailers make more money from farmers who do not grow GMO crops because those farmers need to spend more on pesticides and other inputs. For example, the Colorado potato beetle is enemy number one for potato crops in North America and Europe. Potato farmers who do not plant GMO potatoes have to spray their crops with pesticides seven or eight times per growing season to combat these tiny beasts or risk losing the whole of that season's crop to hungry potato beetles.

Furthermore, the most common GMO potatoes are Bt potatoes. They use a gene from the same bacterium, *Bacillus thuringiensis*, as Bt corn. I mention this because a very, very small percentage (0.001%) of wild potato plants naturally evolved the same gene in low levels to protect themselves against the Colorado potato beetle.[21] In other words, Mother Nature came up with the same adaptation; we just gave it a big kickstart by propagating and producing it in a lab.

The mutation in the wild potato plant is not an isolated event. The same is true for many other plants in nature. This is because Mother Nature has her own genetic engineers. Take *Agrobacterium tumefaciens*,

---

21 Perlak, Frederick J, et al. "Genetically improved potatoes: protection from damage by Colorado potato beetles." *Plant Molecular Biology*. 1 May 1993. 22 (2): 313–321. doi:10.1007/BF00014938

or *Agrobacterium* for short, which can be found in most soil. You can go into your backyard, dig up the soil, and test it, and you will likely find *Agrobacterium.*

Much like scientists in a lab, *Agrobacterium* finds a host plant and inserts its own DNA into the plant. This causes the plant to form a knob, also known as a "gall." The gall secretes a chemical that the bacteria feed off. Essentially, *Agrobacterium* tricks the plant into feeding and protecting it.

Because *Agrobacterium* has been around for millions of years (we know this through gene sequencing), it has inadvertently picked up the DNA of other organisms along the way, for example, other bacteria or plant hosts like wild tomato or wild carrot. So, when *Agrobacterium* inserts DNA into a new host plant, it sometimes inserts little bits of genetic material from those other organisms, and sometimes, those mutations are enough to produce a new species of plant. Thus, *Agrobacterium* has been genetically modifying plants for millions of years. Did you know that Mother Nature and Agrobacterium inadvertently created the sweet potato? A natural GMO created almost the exact same way that man created GMO canola! We know this because *Agrobacterium* DNA pieces are found in sweet potato DNA.

So even if we quit making GMO foods today, the phenomenon is not going to disappear in evolution. Changing genetic sequences is a natural process, and it is happening all around us all the time. Why not learn from Mother Nature and use this process to benefit food production?

Let's do a quick recap. We know that humans have been selective and cross-breeding plants for millennia, and GMO foods are the newest, safest, most reliable iteration of this process. They allow us to grow more resilient, more nutritious crops with less impact on the environment. They lower costs to producers and consumers. They enable farmers to be more sustainable and protect the health of their soil for future generations.

Most significantly, GMO foods are as safe as non-GMO foods according to a substantial body of evidence and overwhelming scientific consensus, including endorsement from the American Medical Association, the National Academy of Sciences, the American Association for the Advancement of Science, and the World Health Organization.

You'll notice that most GMO foods available are staple crops like corn, soybeans, and cotton. That's because the main purpose of developing a GMO crop is to mitigate crop loss. We need more of these crops than others because people rely more heavily on them, and we want to reduce the amount of crops we lose to insects, weeds, and diseases.

Famine and hunger are daily realities or within living memory for millions of people around the globe. Food security is a real issue. Farmers are working hard to combat these issues and provide safe, secure, nutritious food for a burgeoning global population. GMO foods are invaluable tools in that fight. In the next chapter, we'll look closer at how GMO foods can help feed the hungry and end starvation.

Chapter 4

HOW CAN WE END STARVATION?

H.G. Wells wrote, "Civilization is in a race between education and catastrophe." We are witnessing that prescient statement play out before our eyes in the discussion around GMO foods. Certain GMO foods, like Golden Rice, have become a battleground between education and catastrophe, and the stakes are literally life and death.

Golden Rice is genetically modified to significantly increase the beta carotene content, a richly pigmented vitamin found in foods like carrots, pumpkins, and sweet potatoes. The beta carotene in Golden Rice is identical to the beta carotene found in non-GMO plants, and one cup of Golden Rice can supply between 30 and 50 percent of the recommended daily dietary requirement for vitamin A.

If it were allowed to be produced in developing countries, Golden Rice could prevent more than three million childhood starvation deaths annually.[22] But anti-GMO forces have successfully prevented its introduction to growers and consumers. GMO food has never harmed anyone, but disallowing people access to GMO foods like Golden Rice is killing millions.

In addition to supplying desperately needed nutrition, GMOs can also help prevent crop loss and resulting famine and reduce reliance on pesticides. In this chapter, we'll talk about what GMOs could be doing to save the lives of the bottom billion.

---

22 Keats EC, et al. "Improved micronutrient status and health outcomes in low-and middle-income countries following large-scale fortification: Evidence from a systematic review and meta-analysis." *The American Journal of Clinical Nutrition*. 2019;109(6):1696-1708.

A central theme of this book is the ethical imperative we farmers have: to feed people. I feel very strongly about this. It gives my work purpose and drives me to do the best I can for people and the planet, and it is a primary reason I am such a strong proponent of GMO foods.

Using nature's own methods to improve upon our food sources, GMO foods provide more nutritious, higher yielding, disease-resistant crops. This is a winning combination in the fight against hunger and malnutrition. So, what GMO foods have been developed to help end starvation and disease, and why haven't they been more widely adopted?

Let's look at Golden Rice. The World Health Organization estimates that 250 million people worldwide suffer from vitamin A deficiency, and most of those afflicted are 190 million preschool age children and 19 million pregnant women.[23] It may not sound serious, but vitamin A deficiency kills more children than malaria, HIV/AIDS, or tuberculosis.

Vitamin A deficiency is also the leading cause of preventable blindness in children. Each year, this deficiency causes blindness in half a million children, half of whom die within one year of losing their sight. Estimates suggest that making vitamin A available in the daily diet could prevent up to 2.5 million deaths of babies and preschool-age children in developing countries where the risk is highest.[24]

There have been other policy interventions to introduce vitamin A into high-risk populations, such as providing supplements, food fortification (adding vitamin A to other foods), and diet diversification to include more vitamin A-rich foods. However, none of these have been shown to be sustainable or even realistic in remote, impoverished areas where the risk is highest.

In an ideal world, people would have access to a diverse variety of foods that provide the full range of daily nutrition. However, seasonality, crop

---

23 "Golden Rice FAQs." International Rice Research Institute. 2018. https://www.irri.org/golden-rice-faqs

24 Ibid.

loss due to disease, drought, and pests, accessibility, and affordability make the ideal impossible in the real world.

Golden Rice, however, provides a true solution—reliable, sustainable access to a nutrient-dense staple crop—that could be readily implemented in even the most remote communities. After all, rice is already a staple crop in many of the areas most affected by vitamin A deficiency, like sub-Saharan Africa and Southeast Asia!

Golden Rice cultivation practices are the same as for other varieties of rice, so farmers would not need to change how they grow crops or manage their farms. It is a one-to-one substitution in this sense—with the added benefit that Golden Rice has the potential to save lives. Moreover, Golden Rice can work in conjunction with existing interventions like education, dietary diversity, and food fortification.

So, we have a realistic, sustainable solution to a devastating problem that kills millions of children every year. Why isn't everyone planting Golden Rice? The answer is disheartening. Since it first made headlines for passing its "proof of concept" stage in 1999, there were the expected delays caused by the scientific and technological intricacies of engineering a plant with the desired trait.

The current Golden Rice plant is actually a second-generation version, boasting even higher nutritional content; the first iteration was rejected for not having enough beta carotene content to properly address the average dietary deficiency. And of course, rice plants only grow so quickly, so the science had to progress at the pace of the plant.

But the real reason Golden Rice continues to face backlash is the staunch opposition of third-party organizations, including Greenpeace, which oppose all GMO foods in a blanket ban that openly ignores scientific consensus. Groups like Greenpeace have drummed up enough fear and anxiety around GMOs that even the people who would benefit the most from growing Golden Rice are resistant to adopting the new technology.

Greenpeace opposes GMOs in large part due to a distrust of any corporation and a misunderstanding of the science behind genetic engineering, the safety of which we covered in depth in the previous chapter. And perhaps this is also a failing of scientific communication about GMO foods.

The other argument from the opposition is that Golden Rice, and indeed any GMO, will negatively impact biodiversity in the region. But the same argument was made against Bt corn, which now accounts for roughly 80 percent of corn grown in North America. However, there has been no such negative impact on biodiversity since it was first planted in 1996. Indeed, a comprehensive study after one billion acres of Bt corn had been planted revealed that there have been "no unintended or adverse effects" to non-targeted species—including pollinators like honeybees—meaning that the biodiversity in those one billion acres and the surrounding areas was completely unaffected and, in fact, is still flourishing.[25]

In 2016, more than 150 Nobel laureates signed an open letter confronting Greenpeace for its campaign against Golden Rice and other GMO crops. The laureates asserted that Greenpeace had "misrepresented the risks, benefits, and impacts" of genetically altered food plants and, "There has never been a single confirmed case of a negative health outcome for humans or animals from their consumption."[26] That is a resounding endorsement of Golden Rice and GMOs more generally from the top experts in the field.

Even Patrick Moore, the co-founder of Greenpeace, now whole-heartedly promotes allowing Golden Rice after deepening his understanding of the science and the enormously positive potential impact on children's health.[27] He realized that his misguided idealism and advocacy against GMOs were actually harming people and the planet.

---

25 Romeis, J., et al. "Genetically engineered crops help support conservation biological control." *Biological Control.* 130 (2019): 136-154, https://doi.org/10.1016/j.biocontrol.2018.10.001.

26 McKie, Robert. "'A catastrophe': Greenpeace blocks planting of life-saving golden rice." *The Guardian.* 25 May 2024. https://www.msn.com/en-gb/news/uknews/a-catastrophe-greenpeace-blocks-planting-of-lifesaving-golden-rice/ar-BB1n2irz

27 "Allow Golden Rice Now!" Presentation by Patrick Moore, PhD. 22 November 2013. https://www.youtube.com/watch?v=8MCtVqmCoI8

It is also important to note that Golden Rice was developed for humanitarian purposes. Companies including Monsanto granted free licenses to subsistence farmers in developing countries, which disproves claims that farmers would be beholden to corporations or that corporations developed these plants solely for profit. Farmers could grow up to $10,000 worth of crops without having to pay royalties, and they would be permitted to keep and replant seeds for subsequent harvests.

None of the companies involved in its development would receive royalties or payments for marketing or selling Golden Rice—despite spending millions of dollars and decades of time on research and development.

I find it heartbreaking that anyone would deny people access to a lifesaving crop out of fear and deliberate ignorance of the considerable body of evidence that shows GMO foods are safe. Children are starving, going blind, and dying, and we have a solution to their suffering just sitting on a shelf. The people who need Golden Rice the most don't have the power to ask for it, and the affluent people behind Greenpeace who don't need it are the ones preventing people from growing it.

The power of GMO foods to end hunger, malnutrition, and starvation goes beyond Golden Rice. Another facet of GMO foods is the development of disease and pest-resistant crops, as we discussed in the previous chapter.

Now, imagine you are a small farmer in a poor, rural community. You grow food to feed your family, and if there's a surplus, you sell it at the local market for a little profit so you can buy other food, fix your roof, or pay for your children's schoolbooks. Being a small-scale farmer of modest means, you can't afford pesticides or herbicides. You plant corn and hope against hope that corn borers and other pests won't destroy your whole crop before the harvest.

Farmers in Africa lose anywhere from 21 to 82 percent of their crops annually from pests like the corn borer and fall armyworm[28] and parasitic

---

28 Benjamin, J., et al. "Cereal production in Africa: the threat of certain pests and weeds in a changing climate—a review." *Agriculture & Food Security.* 13, 18 (2024). https://doi.org/10.1186/s40066-024-00470-8

weeds can cause 100 percent losses,[29] and such problems are likely to become more severe due to changing climate conditions like increases in drought, flooding, and pest populations.

GMO crops like Bt corn have an innate immunity to pests thanks to a simple change to one gene. Planting GMO corn would eliminate the cost of pesticides and the risk of total crop loss. In a recent study analyzing 40 years of data on Bt corn, researchers found that corn borers were suppressed by more than 90 percent.[30] That is an incredible accomplishment. That level of improved yield translates to a lot of food for a lot of people. Even more impressive is the fact that non-GMO crops in fields neighboring Bt corn also benefited from decreased pest presence.

Furthermore, as we learned in the previous chapter, the advances in GMO crops often allow farmers to adopt regenerative agriculture practices like zero tillage, which improves the health of soil and increases organic matter content. That is critical for regions like Africa where the soil quality is often poor and farmers cannot access or afford fertilizers to replenish the nutrients in soil.

But because of foreign influences dictating what is right or wrong for them, farmers in Africa and other at-risk regions are forced to continue eking out a living without the help of the technology we are already benefitting from in the West. Subsistence farmers are spending all their time trying to grow a little bit of food when they could be growing a lot of food and drastically improving their livelihoods.

Alleviating hunger and protecting crop yield is a great tool for lifting people out of poverty. Those of us in North America are lucky. As mentioned before, just two percent of the population in our region is involved in agriculture. That means everyone else has the time and freedom to

---

29 Ibid.

30 Dively, G., et al. "40 years of data show Bt corn significantly reduces pests, spraying and crop damage, including in nearby non-GMO fields." *Proceedings of the National Academy of Sciences.* 13 March 2018. https://geneticliteracyproject.org/2018/03/13/40-years-of-data-show-bt-corn-significantly-reduces-pests-spraying-and-crop-damage-including-in-nearby-non-gmo-fields/

pursue other interests and careers. We can be doctors, teachers, engineers, innovators—anything we want.

While it is a net positive that so many people are free to follow their dreams and help create the rich, diverse fabric of society, it also means that most people are more removed from the challenges of feeding a nation and the world. We forget that growing food is a full-time job and not everyone gets enough to eat in places with high food insecurity.

For most of human history, most of our time was spent fighting against the forces of Mother Nature. We had to collect and chop wood for fires to keep warm. We had to tan hides or weave cotton to make clothes. We had to hunt, forage, and gather for food. When agriculture began, our fight shifted from big game to small pests that destroyed our crops. We battled drought and floods. A long winter or an unusually dry summer could mean the difference between eating and not eating. Given the ease of modern life, it is hard for most of us to imagine that those battles continue in many places today.

When it comes to GMO foods, the cat is out of the bag, and there is no putting it back. Frankly, I think that is a good thing. Hundreds of research projects covering a period of more than 25 years and involving more than 500 independent research groups have confirmed that GMOs pose no greater risk than conventional plant breeding technology.[31]

Our food is safer than ever, in no small part thanks to GMO foods, which were developed with good intentions to solve real-world problems. We are at lower risk of crop loss, we can use fewer pesticides or even none, and we can give people access to lifesaving, nutrient-dense food—all while protecting the planet for future generations. It is only because of fearmongering from the few that GMO foods have had slower uptake in the regions that could most benefit from them.

I am not saying that we should only plant GMO crops or that they are a panacea for all the world's problems when it comes to health, diet,

---

31 Bruins, Marcel. "You Are Myth Taken: GMOs." *Seed World.* 22 February 2023. https://www.seedworld.com/europe/2023/02/22/you-are-myth-taken-gmos/

and food security, but GMOs are an essential tool in a science-based, multi-pronged approach to meeting the tremendous challenge of feeding nine billion people by 2050. If we want to reach that goal, we will need a combination of conventional, organic, and biotech agriculture.[32]

GMO foods like Golden Rice and Bt corn easily pass the "three-legged stool" test. They add value to society, the farmers who grow them, and the companies that develop them. Most importantly, GMO foods help feed and nourish people who desperately need them. As we continue prove the value of GMO foods to society, I sincerely hope we open our eyes to the evidence and allow the most vulnerable among us access to the same technology and benefits we have already gained from planting GMO crops.

---

32 Ibid.

Chapter 5

ORGANIC – IS IT WORTH THE EXTRA COST?

Every educated consumer knows that organic costs more because it's healthier, but is it really? Is a cookie made with organic ingredients better for you than a conventionally grown apple? Does your organic produce really come from a cute little 20-acre farm down the road from your Whole Foods store, allowing you to feel proud and happy because you are eating local? Despite all the research and studies, there is no clear evidence that organic food is healthier.[33]

Before we dive in, I want to be clear that organic versus conventional farming is not an either/or issue. The two methods are not diametrically opposed, nor do we need to choose one or the other. In fact, farmers who employ both methods can learn from one another.

Here's how I see the issue: organic and conventional agriculture food producers want the same thing. We want to grow nourishing food for people, we just go about achieving our shared goal in different ways. Both systems have their pros and cons, and it is important we understand them.

Let's start with the basics. What is organic food? What standards does a grower have to meet if they want that label on their products? For produce and grains, organic labelling requires they be grown with natural fertilizers and eco-friendly pest-control, and farmers must take certain

---

33 "Are organics worth it?" Harvard Health Publishing, Harvard Medical School. 10 December 2019. https://www.health.harvard.edu/healthbeat/are-organics-worth-it. See also Smith-Spangler C, et al. "Are Organic Foods Safer or Healthier Than Conventional Alternatives?" *Annals of internal medicine.* 2012. 157. 348-66. 10.7326/0003-4819-157-5-201209040-00007.

steps to protect soils and water. As for meat, dairy, and eggs, animals must be allowed to roam freely, and farmers cannot use growth hormones or antibiotics to treat them.

That all sounds grand on the surface. But what does the reality look like? Organic farmers face the same obstacles as conventional growers. They must still deal with pests, fungi, disease, soil management, and other challenges to growth. So, how do they manage? As it turns out, they also use fertilizers, pesticides, insecticides, and fungicides. There are over 100 inputs that are approved for use in North America and Europe.

If we take the first requirement—natural fertilizers—we run into our first organic growing pitfall. We touched briefly on the necessity of nitrogen for plant growth and health in Chapter 2, but plants need four essential nutrients: nitrogen, phosphate, potash, and sulfur. These four nutrients must be present in soil otherwise crops will be nutrient-deficient and won't grow properly. Moreover, crops grown in nutrient-deficient soil are less nutritious when consumed and they grow less often. Many plants, like broccoli, lettuce, and asparagus, can be picked multiple times in one growing season, and how quickly the plant rejuvenates depends on the nutrient supply in the soil.

Thus, fertilizer use is a major issue in organic farming because organic growers cannot replace all the nutrients they take out of the soil. When a crop is grown and harvested, the soil is essentially mined of nutrition. Therefore, over time, the soil degrades, and this creates a long-term imbalance in the environment—unless we intervene.

Roughly 95 percent of farms in America are family-owned.[34] They are passed down from generation to generation, and that means conventional farmers have a vested interest in maintaining the health, productivity, and longevity of their soil.

So, how do you replace the nutrients? Many organic farmers use compost or add animal manure to their topsoil, but there simply aren't

---

34 USDA Census of Agriculture, 2022. https://www.nass.usda.gov/Newsroom/2024/
02-13-2024.php

enough animals in the world to cover every acre of arable land, nor do we have enough food and yard waste or dead plant matter to make that much compost. Therefore, manure and compost are not enough to replace the soil nutrients lost to the harvest.

Some plants, like peas, beans, and clover, are "nitrogen fixers," meaning they increase the level of nitrogen in the soil by converting atmospheric nitrogen to plant-available nitrogen, which allows these plants to provide their own nitrogen. When these plants are plowed down and left on the field to add nutrients and organic matter to the soil, this is called "green manure." However, using nitrogen fixers in lieu of nitrogen fertilizer is inefficient because those plants also take up nitrogen, and farmers have to sacrifice a year of production on that plot.

Thus, these are not feasible solutions to soil degradation on a practical level. Indeed, the science tells us that abandoning nitrogen fertilizers can lead to nutrient undersupply, meaning that even if organic farming could meet the yields of conventional farming—which it cannot—the food it produced would lack the same nutrient density.[35]

The fact remains that whether you are an organic or non-organic farmer, you must replace or replenish the soil of what you mined from it when you harvested the previous crop. According to Dr. John D. Ikerd, Professor Emeritus of Agricultural Economics, "An agriculture that uses up or degrades its natural resource base, or pollutes the natural environment, eventually will lose its ability to produce. It's not sustainable."[36]

Organic production is, therefore, unsustainable because unless we can replace all the nutrition we remove from the soil when we harvest a crop, we are creating a deficit and making it harder to grow crops the following year.

Think of it like a bank account. If you start out with ten dollars and then withdraw eight dollars, what do you need to do to get back to your

35 Connor, D. J. Organic agriculture cannot feed the world. *Field Crops Res.* 106, 187–190 (2008).
36 Ikerd, John D. "What Is Sustainable Agriculture?" Sustainable Agriculture Research and Education. https://western.sare.org/about/what-is-sustainable-agriculture/

original balance? Conventional farmers top up our soil bank accounts with synthetic fertilizers (where eventually, the organic farmer is going to end up broke).

You might be asking yourself, if nitrogen is one of the most common elements in the universe, why isn't it considered a natural fertilizer? Well, on Earth, nitrogen mostly exists in an atmospheric gas form (N_2) that plants can't use.

To use it as fertilizer, atmospheric nitrogen is combined with hydrogen (H_2) under high pressure and gets converted into ammonia (NH_3) or other nitrates. This is what's known as the Haber Bosch process, and it makes nitrogen available to plants in solid or liquid form to use for building proteins and producing chlorophyll. But technically, since the nitrogen is subjected to a chemical reaction, the resulting product is considered synthetic.

Ironically, many organic farmers are using synthetic nitrogen whether they know it or not because lightning can actually produce nitrates much as the Haber Bosch process does. The energy in lightning bolts is powerful enough to split nitrogen atoms into nitrogen dioxide, which dissolves in raindrops from the lightning cloud and falls to the ground as nitrates that get absorbed by plants. These nitrates are identical to the nitrates in the nitrogen fertilizers we buy and sell and put on our land in conventional farming.

But a plant doesn't care if it is using nitrogen from cow manure, lightning, or synthetic urea. To the plant, nitrogen is nitrogen. Plain and simple. And ultimately, you can't tell the difference between organic lettuce and conventional lettuce by testing the nitrogen in it. It would look the same.

As for natural fertilizers and pesticides, it might surprise you to learn that organic farmers can end up inadvertently using a whole host of supposed "nasty" chemicals, from arsenic, which can naturally occur in some rock fertilizers, to strychnine, ozone gas, and hydrogen peroxide.[37] That is because, as in conventional farming, these substances are controlled in

---

37 National List of Allowed and Prohibited Substances. Code of Federal Regulations. https://
www.ecfr.gov/current/title-7/subtitle-B/chapter-I/subchapter-M/part-205/subpart-G/subject-
group-ECFR0ebc5d139b750cd

the methods and quantities in which they can be used, rendering them harmless for human consumption. Still, organic farmers are allowed to use certain chemicals on their crops that the rest of us can't. You can find more heavy metals in organic food than non-organic. You can also find higher levels of lead in organic protein powder than non-organic protein powder.[38] It's important to understand that these heavy metals are still deemed safe for human consumption, but that just because something is organic, it can still contain these elements. For example, in a test of 134 top-selling protein powders, 75 percent of organic plant-based protein powders had measurable levels of lead, while certified organic protein powders had, on average, double the amount of lead of their uncertified organic competitors.[39] Likewise, a 2007 study of Belgian wheat crops found that organic wheat had more than double the amount of lead as conventionally grown wheat. And a 2024 study found that organic chocolate was more likely to contain higher levels of lead and cadmium than non-organic chocolate. I assume it came from a country with poorer regulation on organics than the US. Keep in mind these levels may still be below the maximum MRL[40] and are still safe. I am only making a point.

So, is it safer to eat organic food? Sorry to burst your bubble, but the answer is no. It might be equally as safe. Again, there is no evidence of definitive benefits to an organic diet,[41] but you will certainly pay for the privilege.

Let's move onto the next requirement of organic labelling: eco-friendly pest control. Here's where things get interesting. Like fertilizers, some pesticides are also approved for use in both organic and non-organic

---

38 Geng, Lisa. "Organic Food and Heavy Metals." *Pursuit of Research*. 26 August 2018. https://pursuitofresearch.org/2018/08/26/organic-food-and-heavy-metals/#:~:text=More%20than%2075%20percent%20had%20measurable%20levels%20of,twice%20the%20heavy%20metals%20of%20their%20uncertified%20counterparts.

39 Ibid.

40 Hands, J., et al. "A multi-year heavy metal analysis of 72 dark chocolate and cocoa products in the USA." *Frontiers in Nutrition*. 31 July 2024; 11. https://doi.org/10.3389/fnut.2024.1366231

41 Vigar V., et al. "A Systematic Review of Organic Versus Conventional Food Consumption: Is There a Measurable Benefit on Human Health?" *Nutrients*. 2019 Dec 18;12(1):7. doi: 10.3390/nu12010007.

types of farming. Farmers use these to protect the crops from bugs and diseases. When crops are protected, we get improved yields and ensure the products that consumers buy at the grocery store are in the best condition.

But if organic and conventional farmers can and do use the same pesticides, what is the point of the organic label?

No pesticides authorized for use on food today are genotoxic (meaning they can harm DNA or cause mutations in cells). There are none. And that is an internationally regulated standard. Now, 30 or 40 years ago, before we had good legislation and excellent food safety standards, that may not have been the case, but it certainly is today.

And so, it is also true that there can be some pesticide residue left on food products—in both organic and conventional foods. But that does not need to set off alarm bells in your head.

The dietary intake of pesticides categorically does not pose a risk to health as long as the individual pesticide concentrations in foods are below the maximum residue level (MRL).[42] Food industry regulations are stringent; therefore, foods are strictly monitored and tested at every stage for such residues to ensure they're safe for consumers.

Moreover, when conventional farmers use pesticides that could end up in the food chain, we have to use them pre-harvest, and we must follow very specific instructions for pre-harvest interval use rates. We use exactly the right amount to do the job, and the odds of that amount of product ending up in the food system are zero to nil. Again, we can think of it like taking Tylenol. It's safe in the prescribed dose and following the instructions for use.

Furthermore, as we learned in the previous chapters, the increase in availability of GMOs also means that conventional farmers can use fewer pesticides since many GMOs are bred to repel pests, fight diseases, and resist drought on their own. So, some conventional farmers may actually use fewer pesticides than organic growers!

---

42 Ibid.

In this sense, the rules of organic farming seem arbitrary. Is it more "natural" to use pesticides or to breed a plant that has natural defenses against pests? It would be a real boon to organic growers to be permitted to use GMO crops because then, crops could naturally protect themselves against insects and pests instead of relying on pesticides as they currently do. And here, we can see where "organic" becomes more of a marketing strategy than a practical one.

On top of that, organic farmers get three to four times lower yields for crops in their fields than non-organic farmers. So, if the whole world went organic, we would have three to four times less food with which to feed humanity. Where's the wisdom in that?

Yield is important for a couple of reasons. One, as we've just stated, is that a farmer's job is to feed people. And it's a big job with seven billion people and counting on the planet.

Organic farming produces 30 to 40 percent less yield than conventional farming. And depending on the crop, that can be as high as 50 or 60 percent less. So, for example, a conventional farmer might grow 60 to 100 bushels of wheat, whereas an organic farmer may grow only 5 to 30 bushels. There is no way we could possibly feed everyone on the planet if all our food was produced organically.

The second is that, yes, farming is a profession. Maybe it sounds ugly to say it, but farmers need to make enough money to keep a roof over their heads, feed their families, put gas in the car, pay their bills, and maybe take a little vacation once in a while, just like everyone else. And of course, if a farmer goes out of business because they can't afford to continue operating, that means a loss in the food supply.

There is a kind of running joke in agriculture that no one has ever met a wealthy organic farmer. And that's because when you cut your yield, you cut your profits—or as is the case with organic farming, you need to drastically increase the cost of your goods. That is another reason why most organic farmers partner with big corporations—they have to. No one else can afford to buy their crops and sell them to a consumer base with the means to pay four times the price for essentially the same product.

That choice falls to the consumer. If you prefer organic and your budget can handle it, that's fine. But if what you're concerned about is your health, the science shows that the most important step you can take toward a healthier diet is simply eating more fruits and vegetables, and whether they're organic or conventionally grown does not make a difference.[43]

The reality is that most consumers could not afford to switch to organic. There are people who can barely afford conventionally grown foods, let alone organic, and the problem becomes even more severe as we pan out globally to areas with acute poverty and food scarcity.

Beyond the discussion of food safety and health, there is a debate about the carbon footprint of food production. You might think that organic farmers are more likely to have a smaller carbon footprint because of their methods of production. That is true at the level of total greenhouse gases produced, but only because they produce less food.

However, a better metric for measuring our carbon footprint is to contextualize the numbers in terms of per bushel or per unit of food produced. Organic farmers produce more greenhouse gases and use more water and land per unit of food.

One measuring stick we use in conventional agriculture is water use efficiency. How many inches of water does it take to grow a bushel of crop? We want to grow as much food per inch of water as possible. The same goes for land use. These are limited resources, so we need to use them efficiently.

Another argument launched against conventional farming is that synthetic fertilizers can end up in rivers and streams. While that is possible, it does not reflect the reality of how farmers apply fertilizers today. We have learned from past mistakes and taken steps to correct them.

---

43 "Are organics worth it?" Harvard Health Publishing, Harvard Medical School. 10 December 2019. https://www.health.harvard.edu/healthbeat/are-organics-worth-it

Fertilizers can only end up in waterways if they are used improperly. So, what are we doing differently now?

We soil test, and we put on only exactly how much the plants will need when they need it. We use a shorthand for this called 4Rs: right fertilizer, right rate, right time, and right place—and as minimally as is possible. To prevent the nitrogen we put on our soil running off and leaching into rivers and streams, we put the fertilizer down in a narrow band just below the seed where the roots can intercept it. The result is that the crop gets the fertilizer it needs, and there is just enough in just the right place that only the crop gets it.

On top of simply wanting to do the right thing for people and the environment, it makes sense to use fertilizers and pesticides this way from a financial standpoint. Inputs like fertilizers and pesticides are often the most expensive part of a farm's budget. We only use them when we absolutely have to, and we only use exactly the right amount, so we don't overspend and go bankrupt trying to grow the crop.

Can leaching occur? In Western Canada, where my family farm is located, we don't get enough rainfall to cause leaching, but we are cognizant of these issues, and we try to manage them. We care as much as organic growers about food safety and the health of our soil. Are we perfect? No, but we are constantly trying to improve. The number one focus of any farmer nowadays, conventional or organic, is to increase the nutrient content of the soil and not reduce it. It is a common goal for all farmers.

As for the cute little 20-acre farms of the world, I've got more bad news. The biggest purveyors of organic food couldn't possibly rely on small farms. Organic food is a multi-billion-dollar business dominated by Costco, Whole Foods, and other giants. In fact, 80 percent of all US organic food sales are now made by corporate conglomerates like ConAgra, H.J. Heinz, and Kellogg. The biggest retailers of organic foods are Walmart, Costco, and Kroger. As of 2020, it was a $56 billion industry in the US and growing; globally, that figure was over $134 billion.

And do you know what comes along with corporate conglomerates? Big marketing budgets. It benefits these corporations to sell you on the idea of organic food as healthier and safer because your fear and anxiety—your honest desire to do the right thing for your health, your family, and the environment—translates directly into increased shareholder profits for them.

Many people also assume that organic tends to mean local. You might buy a carton of plums labeled "Jim's Organic Plums" from Whole Foods and picture the lovely little orchard you pass on your way out of town. But more often than not, you will find that those plums come from Mexico or another country with fewer regulations and less oversight than North America. One hundred and ninety countries worldwide participate in organic farming, but only 76 have implemented organic regulations.[44]

Jim's plum farm might not even be a single farm. It could be nothing more than a label created by one guy who has contracts to buy all the plums from a big area and calls his output 'Jim's Organic Plums.'

So, are Jim's plums organic by definition? Sure. Was pesticide used? Probably. Are they safer? There is nothing—no evidence—to confirm that an organic plum is any safer than a conventionally grown plum.

Lastly, we've given organic farming the benefit of the doubt in this discussion and looked at it as if all organic farms are following all the guidelines to a tee. The reality is that organic farming practices can't be generalized and used universally.[45]

No two organic farms are the same. There is no single standardization across the industry, and the number of regulatory bodies that inspect organic farms has yet to catch up with the number of organic farms. So, it is possible that an organic label isn't worth the paper it's written on.

---

44 Willer H, et al. "The World of organic agriculture – Statistics & Emerging Trends 2022." European Commission. 2022. https://knowledge4policy.ec.europa.eu/publication/world-organic-agriculture-statistics-emerging-trends-2022_en

45 Rigby, D, and D Cáceres. "Organic Farming and the Sustainability of Agricultural Systems." *Agricultural Systems*, vol. 68, no. 1, 2001, pp. 21–40., doi:10.1016/s0308-521x(00)00060-3.

I'd like to return to what Dr. Ikerd said about sustainable agriculture. He said, "A sustainable agriculture must be economically viable, socially responsible, and ecologically sound … and the three must be in harmony."[46]

One of my best friends converted his farm to organic production about 20 years ago. He is an educated man, and he decided he was going to try organic farming. He really struggled, but eventually, he figured out how to grow a crop with fewer weeds, and even then, he was only growing 20 bushels of wheat instead of the 60 he used to produce. It was a massive drop in yield. He eked out just enough to get by, but he was never able to advance his farm. I don't think he's even planning to pass it on to his children.

So, did he accomplish his goal? In some ways, yes. He survived, and he turned his farm into a working organic farm.

But was it sustainable? The answer to that is less clear. If we use Ikerd's criteria, the answer is no. It was not economically viable; he suffered huge losses to his former yields and hence, his profits. It wasn't the socially responsible choice because it meant he was producing food for only a third of the people he formally fed and increasing the cost of the food he did produce. And it wasn't necessarily ecologically sound because, as we now know, organic farming eventually depletes the nutrients in the soil. So even if he wanted to leave the farm to his children, he would be leaving behind degraded land.

What I think is most important to remember is that commercial agriculture may not be organic, but it can be sustainable, responsible, and ethical. It more than meets the criteria of being "economically viable, socially responsible, and ecologically sound."

---

46 Ikerd, John D. "What Is Sustainable Agriculture?" Sustainable Agriculture Research and Education. https://western.sare.org/about/what-is-sustainable-agriculture/

Chapter 6

DON'T SOME FOODS CAUSE CANCER?

One of the biggest and most pernicious lies about food is that certain foods cause cancer, particularly pancreatic and non-Hodgkin lymphoma. The media will have you believe these diseases are the result of foods such as canola oil, non-organic foods, GMO technology, and the widespread use of glyphosate in agriculture. Let's get real. Foods grown these ways don't cause cancer. If you eat a Big Mac now and then—I love them—you will be fine. If you eat them every day, it will likely take a toll on your health, but not for the reasons you may think. In this chapter, I'll show you why it's simply not true that food causes cancer.

I want to begin by acknowledging how serious this issue is. Most of us have lost a loved one or friend to cancer, or maybe we know someone who is fighting the good fight as we speak. Anything—*anything*—we can do to stop cancer in its tracks is worth exploring and questioning. But it is equally important with an issue like this, a personal, emotional issue that can strike us right to the core, that we still find a way to listen to the science and the experts on how best to navigate the discussion and what we choose to do when it comes to preventing cancer.

The probability of any food purchased in North America causing cancer as a result of contaminants is almost nil. One reason for this is the food safety and security programs set up by regulatory bodies like the Food and Drug Administration and Health Canada. In the USA, the Food Safety Modernization Act (FSMA) provides strict guidelines and regulations that prevent contamination at different points in the food supply chain.

Now, we won't get too granular on the various laws and organizations involved in regulating and maintaining food safety because in America alone there are over 35 statutes on federal food safety, 28 House and Senate committees providing oversight, four primary government agencies, including the FDA, USDA, and the EPA, and more than 50 interagency agreements between them—and that's just the tip of the iceberg.[47]

And while every country has its own interpretation of food safety standards, each country also consults with higher bodies like the Codex Alimentarius Commission (CAC), a joint program between the World Health Organization and the Food and Agriculture Organization of the UN, when deciding on its particular set of standards. The Organization for Economic Co-operation and Development (OECD) outlines about 150 internationally agreed methods for testing the safety of chemicals and chemical products.[48] The takeaway here is that there are many organizations, and more importantly a whole lot of people in those organizations, whose mission is to keep us and our food safe.

Food safety standards are not implemented or influenced by chemical or pharmaceutical companies nor any one food sector. They are developed by experts in the field based on the most current research available, and they require that food and anything used in food production undergo rigorous review and adhere to standardized safety procedures to ensure there is no sub-contamination of our foods.

So, how exactly is this accomplished? What measures are in place to protect our food on its journey from farm to table?

Let's take foods that are treated with glyphosate in the growing process, like the wheat on my farm and many other farms. Before glyphosate was

---

47 Institute of Medicine (US) and National Research Council (US) Committee to Ensure Safe Food from Production to Consumption. Ensuring Safe Food: From Production to Consumption. Washington (DC): National Academies Press (US); 1998. 2, The Current US Food Safety System. Available from: https://www.ncbi.nlm.nih.gov/books/NBK209121/
48 "OECD Guidelines for the Testing of Chemicals." *OECD iLibrary*. 2024. https://doi.org/10.1787/20745753

introduced to the market, developers had to prove that it was effective at killing weeds without harming humans, non-target species, or the environment.

Then, researchers tested its safety for consumption in a lab by feeding it to mice, rats, rabbits, and dogs and observing the short- and long-term effects. In one study, mice were fed up to 80 ml (almost three espresso shots worth) of glyphosate per day—a dose significantly higher than anything that would ever appear as a trace element in food in an effort to simulate overexposure—for two years, and there was still no carcinogenicity.[49]

Indeed, in 2021, the European Commission released an 11,000-page report analyzing more than 7,000 studies published over ten years on glyphosate's impact on health.[50] It included investigations into the compound's potential to cause genetic mutations, carcinogenicity, reproductive toxicity, specific target organ toxicity (STOT), endocrine-disrupting effects, and environmental impact—basically every possible negative outcome. The report concluded that there is no evidence that the herbicide causes cancer or poses any health risk.[51] Researchers and regulatory agencies around the world have conducted similar scientific reviews and have arrived at the same conclusion.

You might be thinking, *but haven't Monsanto and Bayer paid out billions in lawsuit settlements to people who were exposed to Roundup (glyphosate) and later developed non-Hodgkin lymphoma?"* It is true that the companies lost three trials between 2018 and 2019 that litigated this question but even so,

---

49 Nielsen L. N., et al. "Glyphosate has limited short-term effects on commensal bacterial community composition in the gut environment due to sufficient aromatic amino acid levels." *Environmental Pollution.* 2018; 233:364–376. doi: 10.1016/j.envpol.2017.10.016

50 English, C. "Glyphosate Doesn't Cause Cancer: New EU Report Confirms What We Already Knew." *American Council on Science and Health.* 17 June 2021. https://www.acsh.org/news/2021/06/17/glyphosate-doesnt-cause-cancer-new-eu-report-confirms-what-we-already-knew-156122. See also Boffetta P, et al. "Exposure to glyphosate and risk of non-Hodgkin lymphoma: an updated meta-analysis." *La Medicina del Lavoro.* 2021 Jun 15;112(3):194-199. doi: 10.23749/mdl.v112i3.11123.

51 Goldhaber, S. "Science Finally Winning the Day in Glyphosate Cases." American Council on Science and Health. 11 October 2022. https://www.acsh.org/news/2022/10/11/science-finally-winning-day-glyphosate-cases-16602

the jury found only that Monsanto and Bayer did not do enough to warn users of potential risks, *not* that Roundup caused any individual's cancer.[52]

The juries arrived at their verdicts in large part due to an assertion made by a single organization, the International Agency for Cancer Research (IARC), that glyphosate *may* be carcinogenic.[53] But the IARC's was a lone and contentious finding.

New data and a better understanding of the mechanisms at work in both glyphosate exposure in humans and non-Hodgkin lymphoma have led to acquittals in the five most recent trials on the subject. One expert witness, Dr. Tomasetti, Director of the Center for Cancer Prevention and Early Detection, testified that, "About 90% of the mutations found in non-Hodgkin's lymphoma are attributable to random replication errors. These errors arise spontaneously, and he has asserted that Roundup is not responsible."[54]

The overwhelming majority of the scientific community agrees that glyphosate is clearly, demonstrably non-carcinogenic.

While there is no link between glyphosate and non-Hodgkin lymphoma, there are significant associations between non-Hodgkin lymphoma and tobacco and alcohol consumption, a sedentary lifestyle, obesity, hypertension, diabetes, and high cholesterol.[55]

The Monsanto/Bayer trials also show how important it is to have experts decide these debates rather than courtroom juries or the court of public opinion. If twelve jurors got together and voted that the Earth was flat, would that make the Earth flat? Of course not!

A plant geneticist from the University of Florida put it this way, "Glyphosate continues to survive the intense, periodic scientific scrutiny

---

52 Goldhaber, S. "Science Finally Winning the Day in Glyphosate Cases." *American Council on Science and Health.* 11 October 2022. https://www.acsh.org/news/2022/10/11/science-finally-winning-day-glyphosate-cases-16602

53 Ibid.

54 Ibid.

55 Huang J, et al. "Global burden, risk factors, and trends of non-Hodgkin lymphoma: A worldwide analysis of cancer registries." *Cancer Medicine.* 2024 Mar;13(5):e7056. doi: 10.1002/cam4.7056.

of the world's most rigorous regulatory bodies yet is deemed dangerous in the court of public opinion. Can we please stop letting lawyers and activists make up science that influences farmers' freedom to operate?"[56]

So, that gives us a rough idea of the scope of research done to prove chemicals are safe for use in farming. The next safeguard, as discussed in previous chapters, is that farmers must adhere to strict guidelines on how and when we apply inputs and in what quantity so the compounds have plenty of time to break down long before they enter the food system.

When I prepare to ship the wheat (or any other crop) from my farm, I must sign an affidavit declaring I have followed all protocols to ensure my food does not have contaminants and that no process was used that would cause it to have residues of contaminants that would render the wheat unsafe by exceeding the MRLs.

When it is delivered to the grain buyer, it is inspected for quality like protein, color, mold, and fusarium. Grain companies each have specific protocols to ensure that they are selling safe products to flour mills. Once the flour is produced, it gets tested again. And if the flour is being made into bread, the bread is next in line for testing.

To put it into perspective, I have faith in the food chain system because I follow it all the way through from the seeds on my farm to the bread on my table (and yes, I make my own bread with my very own wheat and rye from time to time!). If I really believed our food would poison me or my family, do you think I would still be eating or serving it?

Seeds, soils, foods, pesticides, herbicides, and everything else involved at any point in food production must face these hurdles to be considered safe for consumers, and the processes and standards are constantly being reviewed and reassessed for best practices.

---

56 English, C. "Glyphosate Doesn't Cause Cancer: New EU Report Confirms What We Already Knew." *American Council on Science and Health.* 17 June 2021. https://www.acsh.org/news/2021/06/17/glyphosate-doesnt-cause-cancer-new-eu-report-confirms-what-we-already-knew-156122

What seems to concern consumers the most, despite the evidence, is that any trace of any chemical might still be found on their food. It is true that trace elements of chemicals are found in foods—both organic and non-organic foods—but that is because the equipment used to test for chemicals is so advanced that we are detecting parts per million (ppm) and parts per billion (ppb). That means in a one-gram sample, you would find one-millionth of a gram or one-billionth of a gram of whatever chemical you are testing for. It is like finding one penny in $10,000 worth of pennies. The chances of such small traces having an ill-effect on our health are vanishingly small, which is exactly why they are allowed.

We have talked about maximum residue limits (MRL), and they play a role here, too. Almost anything could be carcinogenic if you are exposed to enough of it, but MRLs ensure that the amounts that are allowed to show up in our foods are far, far below toxic levels.

For example, mushrooms contain a compound called agaritine, which is considered a carcinogen in mice yet poses no known risk to humans. Now, if you ate 800 pounds of mushrooms a day for several years, maybe it would cause cancer, but you would need a humongous appetite to eat that many mushrooms! In contrast, tossing a handful into your salad or pasta is beneficial to your health because mushrooms are also rich in antioxidants, fiber, vitamins, and minerals.

Take radiation as another example. Most people think of radiation exposure as extremely dangerous, but we can and do safely absorb small amounts of natural radiation every day. Taking a flight, getting an X-ray, or living within 50 miles of a coal plant all expose us to small, absorbable doses of ionizing radiation, as does sleeping next to someone or eating a banana.[57] But these doses are so infinitesimally small—a fraction of a fraction of a fraction of a lethal dose—they have no impact, adverse or otherwise, on our health.

However, a massive dose all at once will make you sick and maybe even lead to death. Essentially, it takes direct exposure to a disaster like

---

57 Munroe, R. "Radiation Dose Chart." https://xkcd.com/radiation/

Fukushima or Chernobyl to cause severe or fatal radiation poisoning. Even then, "Research has suggested that being close to the nuclear bomb when it detonated in Hiroshima would be less of a threat to your health than being severely obese."[58]

The same is true for trace elements of glyphosate, pesticides, and other compounds found in our food. Too much of any one thing can probably kill you. Marathon runners have been known to die from drinking too much water while hydrating.[59] Does that mean you should stop drinking water? This is all to say that anything might cause cancer at very extreme dosages, but the likelihood of that is about the same as being able to eat 800 pounds of mushrooms in the first place.

Finally, if there are trace elements in food that meet the requirements for MRL, most herbicides and chemicals cannot survive a trip through your digestive system. Our stomachs have a low pH, meaning they are highly acidic. This is so the stomach can break down food and absorb nutrients. That is also going to eliminate pretty much all those trace elements. Any toxins that get through are filtered through our liver and kidneys and excreted out.

But if food with unknown contaminants don't cause cancer, what does? Family history and genetic makeup play roles, as do environmental factors, diet, lifestyle, and random mutations in organs and tissues. Health in general is a complex, multifactorial issue. In a review of evidence linking diet and nutrition to cancer risk, researchers have found that alcohol and obesity are the most significant factors.[60] What does this tell us?

---

58 Wighton, K. "From radioactive sweat to airport scanners—top myths debunked by expert." *Imperial College London.* 16 March 2011. https://www.imperial.ac.uk/news/171326/from-radioactive-sweat-airport-scanners-myths/

59 "Too much water can be life-threatening for marathoners." *The Harvard Gazette.* 2 May 2007. https://news.harvard.edu/gazette/story/2007/05/too-much-water-can-be-life-threatening-for-marathoners/

60 Key TJ, et al. "Diet, nutrition, and cancer risk: what do we know and what is the way forward?" *British Medical Journal.* 2020 Mar 5. 368:m511. doi: 10.1136/bmj.m511.

If you follow a good, standard diet of balanced carbohydrates, proteins, and fats with plenty of fruits and vegetables, (and you should talk to a good nutritionist about that) the likelihood of any one food or substance giving you cancer is almost nonexistent—and the reason I don't say "zero" is because anything is possible. We don't know everything, but to the best of our very deep, expansive knowledge, most foods don't cause cancer.

Are you raising an eyebrow at "most foods?" The reason I say "most" is because there is limited evidence that red meat is carcinogenic to humans and sufficient evidence that processed meat (meat that has been cured, salted, fermented, or smoked) is carcinogenic to humans.[61] But again, we need to apply common sense and remember that correlation is not causation.

These findings do not mean that if you enjoy a nice Sunday breakfast of bacon and eggs you will get cancer. It means that if you eat *a lot* of processed meat *every day*, then the chance you *may* get certain types of cancer *may* increase. The current recommendation is around 70g of red or processed meat per day[62]—so if you eat a Big Mac every day (90g), you might consider ordering a regular cheeseburger (45g) instead. That's it.

Furthermore, as we discussed in the chapter on organic foods, most experts agree that there is no evidence that eating organic food provides any significant advantage to your health.[63] Organic and non-organic foods are both equally safe and equally healthy by most standards.

I want to take a moment here to add that not all science is good science. The best science invites us to explore the complexities and nuances of a

---

61 International Agency for Research on Cancer. Red Meat and Processed Meat. IARC Monographs on the Evaluation of Carcinogenic Risks to Humans. 2018. Vol 114. http://publications.iarc.fr/Book-And-Report-Series/Iarc-Monographs-On-The-Identification-Of-Carcinogenic-Hazards-To-Humans/Red-Meat-And-Processed-Meat-2018

62 "Does eating processed and red meat cause cancer?" *Cancer Research UK*. 26 July 2023. https://www.cancerresearchuk.org/about-cancer/causes-of-cancer/diet-and-cancer/does-eating-processed-and-red-meat-cause-cancer#keyrefsmeat0

63 "Are organics worth it?" Harvard Health Publishing, Harvard Medical School. 10 December 2019. https://www.health.harvard.edu/healthbeat/are-organics-worth-it

question, while the worst science pushes us to ignore them. I've referenced many studies in this chapter, as I do throughout the book, and I want to point out that these studies are authoritative and peer-reviewed, and the findings are repeated and validated over and over again by other scientists in other places with other backgrounds and perspectives. In some cases, dozens, hundreds, and even thousands of studies all come to the same conclusion about the safety of our food.

There is also bad science. By that I don't mean that the people conducting such studies are bad people, not at all. But even people with good intentions can end up putting out junk science based on poor methodology, deductive reasoning, broad extrapolations from small, unrepresentative sample sizes, or cherry-picked data.

For example, two authors in particular, Samsel and Seneff, published a series of studies making negative claims about glyphosate that have since been deemed incorrect, deductive, distracting, speculative, unsubstantiated, and inappropriate by their scientific peers, yet their conclusions were in circulation for years and likely picked up by media outlets that further spread their fallacies.[64] And it is hard to put that genie back in the lamp.

There are also people who will try to profit off our vulnerabilities and fears. If you come across any diet that claims it can prevent or even reverse all types of cancer and chronic disease, it might be time to scrutinize those claims.

The homeopathy market size in the US was nearly five billion dollars in 2023, and it is anticipated to skyrocket to over 18 billion in the next ten years.[65] That is a sizeable chunk of change for a broadly unregulated market.

Homeopathic remedies don't need to be tested for safety nor are manufacturers required to prove the efficacy of their products—unlike

---

64 Mesnage, Robin & Antoniou, Michael. "Facts and Fallacies in the Debate on Glyphosate Toxicity." *Frontiers in Public Health.* 2017; 5: 316. 10.3389/fpubh.2017.00316.
65 "Global Homeopathy Market Overview." *Market Research Future.* 2024. https://www.marketresearchfuture.com/reports/homeopathy-market-4970

any product used in food production or any food produced in commercial agriculture. So, it might be wise to remember that if someone is telling you that a particular food is harmful and will cause cancer, that same person is probably trying to sell you a supplement, sexy new diet, or miracle drink they claim will make you healthier.

I was looking at a jar of multivitamins the other day. The label listed potassium and phosphorous—the exact same nutrients I use for plant food. So, a person who eats only organic food and takes multivitamins is essentially consuming the same thing they are trying to avoid in non-organic food—and paying $30 for the privilege!

It isn't just the snake oil salesmen trying to take advantage. Food manufacturers and retailers also stand to make a pretty penny from your anxieties. Organic food, as previously discussed, can cost upwards of four times more than non-organic food, and companies are more than happy to slap that little organic sticker on anything they can if it means increasing their profit margins. I have even seen bottled water labeled "organic." Organic water. I don't know what the heck that is.

The same thing goes for fears about GMO foods. As referenced in previous chapters, we know that GMOs are not linked to any cancers because there are similar cancer rates in the US, where GMO consumption is high, and Europe, where GMO consumption is lower.[66] If there was any link between GMO foods and cancer, we would expect to see higher cancer rates in the US—but, of course, we do not. Nevertheless, you'll pay a dollar more per loaf for non-GMO bread, and guess what? All US bread is non-GMO because there is no GMO wheat.

North Americans are healthier because of the food we have access to. Most of us don't face malnutrition, starvation, or diseases associated with these issues, as people in many other places in the world still do.

---

66 "Genetically Engineered Crops: Experiences and Prospects- New Report." National Academies of Science, Engineering, and Medicine. 17 May 2016. https://www.nationalacademies.org/news/2016/05/genetically-engineered-crops-experiences-and-prospects-new-report

We have an abundant food supply, and most of us get all the nutrients and micronutrients we need.[67]

It is also critical to note that, "Cancer deaths are dropping every year by 1 or 2%. When you add that up over 20 years, cancer deaths are down by almost 25% from where they were at the turn of the century."[68] Diagnoses of and deaths from all types of cancer are declining[69], and it is important to overlay that with the fact that in the same period, our use of GMOs, food additives, and chemicals like glyphosate have increased. If ordinary, commercially produced foods caused cancer, we would expect to see the opposite effect.

Like I said at the beginning of the chapter, most of us have lost someone to cancer or know someone who has. I think that is at the heart of why many people are susceptible to misinformation around diet, food, and health. We want to believe we can protect ourselves and our loved ones from the tragedy of loss.

The good news is that, in some ways, we can because our food *is* safe. To the extent that our diet can help prevent cancer and other diseases, it is primarily as a means of promoting a healthy lifestyle and preventing obesity, which is a precursor for many chronic health issues. But that isn't because food causes cancer.

---

67 Committee on a Framework for Assessing the Health, Environmental, and Social Effects of the Food System; Food and Nutrition Board; Board on Agriculture and Natural Resources; Institute of Medicine; National Research Council; Nesheim MC, Oria M, Yih PT, editors. A Framework for Assessing Effects of the Food System. Washington (DC): National Academies Press (US); 2015 Jun 17. 3, Health Effects of the U.S. Food System. https://www.ncbi.nlm.nih.gov/books/NBK305175/

68 Simmons-Duffin, S. "The NIH director on why Americans aren't getting healthier, despite medical advances." *Houston Public Media.* 7 December 2021. https://www.houstonpublicmedia.org/npr/2021/12/07/1061940326/the-nih-director-on-why-americans-arent-getting-healthier-despite-medical-advances

69 Committee on a Framework for Assessing the Health, Environmental, and Social Effects of the Food System; Food and Nutrition Board; Board on Agriculture and Natural Resources; Institute of Medicine; National Research Council; Nesheim MC, Oria M, Yih PT, editors. A Framework for Assessing Effects of the Food System. Washington (DC): National Academies Press (US); 2015 Jun 17. 3, Health Effects of the U.S. Food System. https://www.ncbi.nlm.nih.gov/books/NBK305175/

The best thing we can do for our health and longevity is eat a balanced diet, get a little exercise on a regular basis, and limit or eliminate our intake of alcohol and tobacco. The answer isn't flashy and it isn't new but it is the truth.

Chapter 7

AREN'T CHEMICALS AND ADDITIVES DANGEROUS?

Imagine there has been a chemical spill. An emulsion of triglycerides, sterols, carotenoids, cerebrosides, butyric acid, and nucleic acid spreads across the ground, seeping into cracks and coating everything it touches. A baby cries. People scramble to contain the damage. Are you worried for your safety? Well, unless you're scared of a little spilled milk, you shouldn't be.

The media and alternative health influencers would have you believe that most chemicals and food additives are approved based upon lobbying and that the safety of these additives is proven only by crooked science. Please. It's not true, and it's not fair. I find that particularly people who live in the US seem to think all food safety protocols were built by someone or some organization in the US. They seem to believe that the only controls are their supposed crooked government, big business, or lobbyists. In fact, safe food is the result of many scientific guidelines analyzed by scientists with protocols in many countries from all over the world. Yes, there is a much bigger world than just the USA! The USA is just a part of a much larger collaboration.

In this chapter, we'll talk about the persistent falsehoods spread about chemicals and additives and trace the evolution of chemicals and additives used to produce food. We'll also discuss how these substances can enhance safety, improve nutritional value, make food tastier, and prolong the shelf life of foods. It's my goal that you'll come away from this chapter recognizing that chemicals and additives actually make food safer.

The received wisdom in many circles today is that chemicals and additives are unhealthy and even dangerous. The belief is that they endanger people, harm the planet, and make the food supply unsafe. Researchers in the field of psychology have even found that "Several studies … revealed that people's perception of the naturalness of an entity is more influenced by its processing history than by its content, and perceived naturalness is reduced by contact with unnatural entities and by chemical transformation, especially with additives."[70] In other words, our perception often matters more to us than the reality.

We also tend to overreact to some dangers. Have you ever eaten something that had gone slightly off or made you sick and then been put off that food indefinitely? A bad batch of mussels from a dodgy food truck doesn't mean all mussels are poison; it means *those specific* mussels weren't prepared properly.

We tend to overestimate certain risks, especially those that become matters of "public concern," garner lots of media attention, or which we have a "recency bias" for, meaning we have heard about this risk more recently than some other factor.[71] In other words, our perception of risk doesn't match the statistical reality of risk, including risks regarding chemicals. So, for example, if you recently saw a post on social media claiming that Subway's 9-grain bread is "toxic" because it contains azodicarbonamide, then you are more likely to be fearful of that risk and avoid eating at Subway than you are of jumping in your car and driving on a multi-lane highway, which carries a much higher statistical probability of risk.[72]

---

70 Li & Chapman. "Why do people like natural? Instrumental and ideational bases for the naturalness preference." *Journal of Applied Psychology*. 2012; 42(12):2859-2878. https://doi.org/10.1111/j.1559-1816.2012.00964.x

71 Paul Slovic et al. "Facts and Fears: Societal Perception of Risk." *Advances in Consumer Research*. 1981; 08: 497-502. https://www.researchgate.net/publication/292458918_Facts_and_Fears_Societal_Perception_of_Risk

72 Kennedy, James. "Chemophobia: How We Became Afraid of Chemicals and What to Do About It." *Kennedy College*. 22 November 2023. file:///C:/Users/laure/Downloads/Chempohobia.pdf

There are a host of reasons why very little of what people believe about chemicals is true, but I want to begin by returning to our opening example. People tend to think of chemicals with a negative connotation, i.e., as a harmful contaminant or undesirable additive. We even have a word for it—chemophobia—meaning the irrational fear of chemicals perceived as synthetic.[73] Taken to the extreme, chemophobia can manifest as multiple chemical sensitivity (MCS), which is defined as a psychiatric disorder.[74]

The reason chemophobia is described as an "irrational" fear is because, well, pretty much everything is a chemical because a chemical is simply a substance with a specific composition, whether it be an element, an alloy, or a compound. In this sense, water is a chemical. Steel, aspirin, and sugar are chemicals. The air is made up of chemicals, and so is your pet dog or cat, if you have one. Even you are composed of chemicals. All matter is a chemical or consists of a chemical. Furthermore, some chemicals are naturally occurring and some are synthetically made. The latter are often baselessly subject to even greater demonization, but, as we know, "natural" does not necessarily equate to "safer," and "synthetic" is not shorthand for "dangerous." As we learned in the chapter on organics, some chemicals can even be produced both naturally *and* synthetically, such as nitrogen, which can be produced by lightning and by the synthetic Haber Bosch process.

There are hazardous chemicals—there is no question about that—and some concerns regarding chemical use stem from genuinely harmful examples in history, like the use of DDT and Agent Orange. There are also bad actors like DuPont, which covered up the harms of Teflon. But it is important to remember that we learned from these early mistakes and implemented safety measures to prevent them from happening again.

Let's look at DDT as an example. Dichlorodiphenyltrichloroethane, or DDT, was originally introduced in the US around 1943 to protect

---

73 Ibid.

74 Rosner, R. and Scott, C. *Principles and practice of forensic psychiatry.* 2003. 2nd ed. CRC Press, pp.304-306.

American soldiers fighting overseas from tropical illnesses like malaria and typhus. It did this by killing mosquitos, fleas, and lice that transmitted diseases. DDT was lauded as a miracle chemical, and because of its efficacy, it was soon made available to consumers and for industrial applications.

However, emerging research quickly put the brakes on DDT use. As early as the 1950s, the US Department of Agriculture began regulating and even prohibiting DDT use in various contexts because of a growing body of evidence regarding its environmental and toxicological effects.[75]

So, why did it take until 1972 to fully ban it in the US? Critics have floated conspiracy theories that research backing DDT use was industry-funded or profit-driven, but the truth is closer to *Oppenheimer* than *Erin Brockovich*; science is a process, not an event.

As one Stanford scientist wrote, "For instance, the Forest Service published studies highlighting the effectiveness of DDT in 1948. Their conclusions were often widely criticized, but their science is actually generally considered sound … there didn't have to be foul play or biased motives involved when one study found that DDT could be harmful to natural ecosystems and another found that it was effective at controlling spruce budworms."[76]

The important lesson we can take away from DDT is that when the dust settled and the science became clear, the then–newly founded Environmental Protection Agency (EPA) acted decisively. DDT use was discontinued entirely in the US and we have come a long way since then (although it is still used in some places, like India, where the threat of malaria outweighs other potential harms).

Our understanding of the environment, the balance of nature, and our impact have evolved tremendously and for the better over the last 50 years. In its first years, the EPA passed the Clean Air and Clean Water

---

75 "DDT – A Brief History and Status." *United States Environmental Protection Agency.* 12 March 2024. https://www.epa.gov/ingredients-used-pesticide-products/ddt-brief-history-and-status

76 Sonnenburg, J. "Shoot to Kill: Control and Controversy in the History of DDT Science." *Stanford Journal of Public Health.* 1 May 2015. https://web.stanford.edu/group/sjph/cgi-bin/sjphsite/shoot-to-kill-control-and-controversy-in-the-history-of-ddt-science/

Acts, restricted the use of lead in paint, phased out leaded gas, and began regulating manufacturing and vehicle emissions. Regulatory bodies in many countries built new regulations, not just the US. Similarly, the EPA carved out a new role for their federal governments in many countries to determine exactly what levels of various pollutants were safe, conduct research, and set national protocols and standards to protect the public and the environment. As a result of lessons learned the hard way in the 20th century, we have built extensive global safety nets that are improved upon and refined year after year.

Today, in many countries of the world, every substance involved in food production must be registered, rigorously tested, and meet safety standards set by the OECD and other regulatory bodies *before* they enter the marketplace. It is actually an exceedingly difficult, years-long (or sometimes decades-long) process to get a new chemical or agricultural technology to market.

When we do use approved chemicals, we use them within specific limits and guidelines. However, humans don't have a well-developed innate understanding of dosage, which is problematic because the dose makes the difference.[77] You can drink a glass of water without any risk, but if you were to drink six liters at once, that could be lethal.[78]

And it is just as important to remember *why* certain chemicals have been developed. The reason will always be to meet a specific need and generally to give us some advantage in the ongoing battle against the forces of nature.

If that is the case, then why, if you Google "harmful food additives," will you find pages and pages of infographics and listicles with images of pizza and potato chips telling you the top ten most harmful foods and food additives? The reason, as we have seen so often in this book,

---

77 Kennedy, James. "Chemophobia: How We Became Afraid of Chemicals and What to Do About It." *Kennedy College.* 22 November 2023. file:///C:/Users/laure/Downloads/Chempohobia.pdf
78 Ibid.

is that someone stands to make money by persuading you that normal, affordable food is unsafe.

MSG is a great example of this. First invented by a Japanese scientist who wanted to mimic the flavor of a popular soup stock, it is a synthetic version of a naturally occurring chemical called glutamic acid. If you have ever eaten aged cheese or heirloom tomatoes, then you have eaten MSG. It has roughly one-third the sodium of regular table salt and just as much flavor.

Like many technologies, it was first adopted by the military to make MREs (meals ready to eat) taste better, and it then spread to consumers. That is, until it came under fire in the 1960s for being "toxic" and causing "Chinese Restaurant Syndrome" consisting of headaches and abdominal pain from consuming MSG, and the negative label stuck.

As it turned out, the original scholarly article that identified "Chinese Restaurant Syndrome" and launched the decades-long skepticism surrounding MSG was written as a prank.[79] It started as a bet between an orthopedic surgeon, Howard Steel, and another doctor, Bill Hanson, who, according to *Colgate* magazine, "used to rib Steel about his specialty, saying orthopedic surgeons were too stupid to get published in a prestigious journal such as the *New England Journal of Medicine*. In fact, he bet Steel $10 he couldn't make it into its pages."[80] Writing under the pseudonym Robert Ho Man Kwok, Steel made the whole thing up.

The article was published, Steel collected his ten dollars, and media outlets leapt at the opportunity to send consumers into a new tailspin panic, and simultaneously sell papers and advertising space. Steel contacted the journal to tell them the research was fake, but they would not listen. The myth gained traction and, to this day, you can find people who claim to get "Chinese Restaurant Syndrome" and articles claiming that MSG is

---

79 Blanding, M. "The Strange Case of Dr. Ho Man Kwok." *Colgate Magazine.* 2 January 2020. https://news.colgate.edu/magazine/2019/02/06/the-strange-case-of-dr-ho-man-kwok/
80 Ibid.

one of the most harmful food additives, despite all the evidence to the contrary. And those same people are probably trying to sell you moon water or celery juice instead.

There is also big money in lawsuits. The tort bar sees an opportunity whenever a new chemical is introduced to market. If they can convince a jury that poor Aunt Martha died because a little bit of Round Up touched her rutabagas, then the lawyers can clean up with class action lawsuits. Meanwhile, they are scaring people away from eating perfectly safe and healthy food.

So, there are good chemicals that we think are bad. What about bad chemicals that we think are good? We use all kinds of chemicals every day without second-guessing them. For example, if you wake up every morning and put on deodorant, there is a high probability that you are exposing yourself to chemicals like aluminum chloralhydrate, which creates a barrier against sweat, or triclosan, an antibacterial used to stop the growth of odor-causing bacteria. In large quantities, aluminum chloralhydrate and triclosan can be very dangerous, and in the early 2000s, concerns arose that compounds like these found in deodorants might be linked to breast cancer.[81]

So then how can it be safe to put these chemicals on one of the most sensitive areas of the body? We know they are safe because those chemicals had to be tested just like every food additive or crop protection product. They all go through the same process, and there are limits to how much can be used in each product, just like there are limits to how much farmers can use on crops. And the claims regarding cancer have long been debunked. Doctor Susan Massick of Ohio State University says, "For a compound to cause cancer, it would have to be absorbed into

---

81 Massick, Susan MD. "Should you worry about aluminum in your antiperspirant?" *Ohio State University.* 9 February 2024. https://health.osu.edu/health/skin-and-body/aluminum-in-antiperspirant

the bloodstream at a concentration high enough to cause toxicity. That's not going to happen with a daily dab of antiperspirant."[82]

It is interesting to note that triclosan is even used in other products, like pesticides and toothpastes, for the same antibacterial benefit. And if it doesn't cause concern when you apply it directly to your skin or brush your teeth, there is no reason it should cause concern being used early in the growing process for foods, where only trace elements, if any, will remain once the food enters the food supply.

The decision to use products like this is ultimately up to you, the consumer, and the choice may boil down to whether the risk (in this case, the exceptionally minute possibility that all the testing and safety regulations have missed some obscure long-term correlation to an unwanted outcome) is worth the benefit (being able to walk around in society without smelling like a dog).

We tend to regard products we use every day as normal and safe—and rightly so. Think about all the cleaning products you have under your kitchen sink or the cosmetic products in your bathroom. Do you use bleach to clean surfaces, detergent to wash your clothing, and dish soap to sanitize your dishes? Those contain chemicals!

Many of the chemicals in your household products are the same or similar to those used in food production. Phosphate is used in dish soap. It is related to the glyphosate I use on my crops and to sodium phosphate, which is used as a food additive. You will even find gardening experts who advise using a mixture of water and dish soap as a homemade pesticide for your garden. Does that mean it's safe to drink dish soap? Let me give you a clue—don't try that one at home, folks. Again, a substance's safety depends on how it is used and in what quantity.

---

82 Ibid. See also, Willhite CC, et al. "Systematic review of potential health risks posed by pharmaceutical, occupational and consumer exposures to metallic and nanoscale aluminum, aluminum oxides, aluminum hydroxide and its soluble salts." *Critical Reviews in Toxicology.* 2014 Oct; 44 Suppl 4:1-80. doi: 10.3109/10408444.2014.934439.

Perhaps you're thinking, *I only use natural, organic cleaners and cosmetics.* While "natural" and "organic" may be great for marketing campaigns, they hold little meaning when it comes to the safety and efficacy of products. As we discussed in the organic food chapter, organic producers are still allowed to use chemicals and chemical additives, so long as they are on the approved list.[83] That means that there can still be trace elements of pesticides and other chemicals on your organic apple or in your organic detergent.

Why is that? Because as we noted at the start of this chapter, pretty much every substance is a chemical, and most chemicals are not inherently "safe" or "dangerous!" That evaluation can only be determined based on context and quantity. So even naturally occurring chemicals—like the agaritine found in mushrooms—may be safe at low levels and toxic at extremely high levels. That is why our regulations—like MRLs for any substance that enters the food supply—are so vitally important. It's all about the dose.

We regularly ingest any number of carcinogens that are naturally occurring in foods, for example, neochlorogenic acid in coffee, 1-phenylethanol in chocolate, ethyl acrylate in pineapples, or 8-methoxypsoralen in celery. However, these are perfectly safe because these chemical compounds are present in levels far below the MRL.[84] There is one last category of chemicals and additives we haven't talked about: substances added to food to improve taste, increase nutritional value, or extend shelf life. Like any substance introduced into the food supply, food additives undergo a thorough safety investigation by an independent, international expert scientific group—the Joint FAO/WHO Expert Committee on Food Additives (JECFA)—before being introduced to the public. Only after

83 "National List of Allowed and Prohibited Substances." *Organic Trade Association.* https://ota.com/advocacy/organic-standards/national-list-allowed-and-prohibited-substances
84 Kennedy, James. "Chemophobia: How We Became Afraid of Chemicals and What to Do About It." *Kennedy College.* 22 November 2023. file:///C:/Users/laure/Downloads/Chempohobia.pdf

a chemical has undergone a JECFA safety assessment and been deemed not to present a health risk to consumers can it enter the market.[85]

Everyday items have food additives that help improve the product for the consumer. Milk is fortified with vitamin D and citric acid is added to Campbell's tomato soup. Citric acid is one of the most widely used additives, and like many of the substances we've discussed, it can be both naturally occurring (as in citrus fruits) and synthetically made. It acts as flavoring and as a preservative by preventing the growth of microorganisms that can cause food-borne illnesses or spoilage. That is a pretty significant benefit in my book.

Would you stop buying oranges because they contained citric acid? Of course not. So why would you avoid other products that use the synthetic version? Both are equally safe.

Our society has developed astonishingly high standards when it comes to food safety. We are more informed than ever about what exactly is in our food, and as a farmer, I rely on that information. When a new product comes out, the first thing I ask is how did it perform under all the requisite tests? Why did it get approved? What benefit does it offer? Is it better than what I currently use?

As a society, we are living longer, healthier lives. Are we perfect? No. Are we getting safer? Every day. Are the requirements today much better than they were in 1940 when DDT was invented? Absolutely, yes, they are. Do we do everything possible to reduce our exposure to potentially harmful chemicals in our food? Yes, we do.

Our technology is so sensitive we can detect trace elements of almost anything. I mean, we can even determine the chemical composition of stars and planets in galaxies light-years away! A trace element of a chemical in our food does not mean the food is unsafe, it simply means some tiny particle of the chemical is present.

---

85 "Food Additives." *World Health Organization.* 16 November 2023. https://www.who.int/news-room/fact-sheets/detail/food-additives

Frankly, this discussion boils down to risk/benefit analysis. We are constantly choosing between the risks and potential benefits of everything we do. Driving a car poses inherent risks, but most of us don't hesitate to jump behind the wheel because we know it will get us to our destination faster, and there are airbags and all sorts of other safety measures built into the car.

Likewise, the evidence is overwhelming that commercially produced food is safe. You can interrogate the use of glyphosate or the millions of other chemicals that enable us to grow higher yielding, more nutritious food simply because they *are* chemicals.

It's important to ask questions and be curious, but it is even more important to accept the answers that science provides—even if it means having to change your mind. Hopefully, with a better understanding of the big picture, you can then make a rational decision as to whether the benefit is greater than the risk, to society or yourself.

At the end of the day, the science is settled. Our food is safe. If you still prefer to avoid the infinitesimally small risk of ingesting one part per billion of a synthetic chemical and instead opt to pay five times more for food with one part per billion of a natural chemical, then by all means, fill your boots (and empty your pockets).

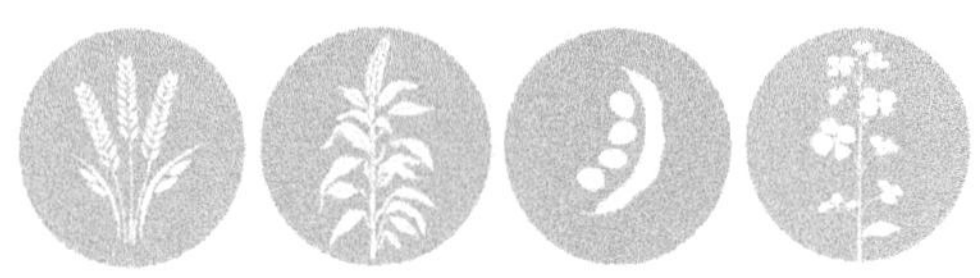

Chapter 8

SHOULD I BE WORRIED ABOUT PRESERVATIVES?

Have you ever opened a bag of potato chips only to discover they tasted just plain awful? Well, that means the oil was rancid. What about going to the movies and getting a nice big bag of popcorn only to find it is completely stale? That happens because food is in a race against time with rot.

Now, imagine instead that you're enjoying a Fourth of July barbecue brisket or a pastrami sandwich or even bacon and eggs on Sunday morning. Those meats have all been preserved by smoking or curing. If you like pickles, sauerkraut, or yogurt, those foods have also been preserved through fermentation. Humans have been preserving food in various ways since the Paleolithic Era.

In harsh climates, the only way humans survived through the ages was by putting up food to be preserved and eaten during winter months or lean harvests. If there were no preservatives, no one would ever have lasted through a single winter or harsh summer drought. In this chapter, I'll demonstrate that preservatives make your food safer and help it last longer. Trust me—scientists in lab coats at food manufacturing companies aren't trying to find ways to make you sick. Preservatives protect the food supply and don't damage your health.

There are dozens of traditional methods of preserving foods, including boiling, canning, curing, fermenting, smoking, jellying, pickling, sugaring, and even burying. These methods were critical to our survival before the dawn of electric refrigeration and other modern preservation methods, and because these methods also add to the flavor of foods, many of them are still in use by home cooks and in mainstream food production in foods like charcuterie, beef jerky, jams and jellies, smoked salmon, pickles, cheeses, beer, wine, and spirits. My wife and I can our own vegetables and make our own sausages at home—and they're delicious.

You might perceive these foods and traditional methods as "more natural" and therefore "healthier" or "safer" … but if you've been reading closely so far, you might be able to predict that those assumptions don't stand up to the science. We live in a modern world, and people cleverer than me have come up with several ways to preserve food without needing to employ the same time-consuming—and, frankly, unpredictable—processes.

With the global population tipping ever faster toward nine billion people in every corner of the planet, we need to be able to transport food without the risk of it rotting before it reaches the family table. To feed everyone, we must be able to preserve food longer, and we've developed a number of methods to achieve that goal without compromising food safety. In fact, our food is safer *because* of these methods and preservatives.

So, what methods and preservatives are most common today? Most people are familiar with pasteurization, vacuum sealing, and aseptic or sterile processing, and that familiarity makes us less concerned. But what about irradiation, pulsed electric field (PEF) electroporation, modified atmosphere, or nonthermal plasma? Do you suddenly feel you've been transported into a science fiction story?

It is important in these conversations to remember that just because we aren't familiar with something or we don't fully understand it, that doesn't mean it's inexplicable or dangerous. As we've discussed throughout the book, anything and everything involved in our food supply system is studied, tested, and retested extensively and thoroughly.

Let's take irradiation as an example. Using Cobalt-60 as a radiation source and shooting electron beams out of an electron gun at your food might give you pause. Like other processing methods, food irradiation causes small chemical changes, which, in turn, produce new substances called "radiolytic products."

You may be thinking, *radiation—isn't that bad?* Well, here is where our old friend MRL comes in. The amount of radiation food is exposed to is closely measured and limited to what is necessary to kill off potentially harmful bacteria and microbes. The radiolytic products are measured in parts per billion and can only be detected with highly sensitive laboratory equipment. More significantly, they are identical or similar to substances that occur naturally in food that is not irradiated, and they have been shown to be harmless in the very small amounts produced.

How do we know? Irradiated food was tested on multiple generations of laboratory animals, and later, on human volunteers, with no ill effects. Reputable national and international organizations, including the American Medical Association and the World Health Organization, endorse irradiated foods as safe. Scientists from the Food and Drug Administration, the US Department of Agriculture, and universities across the globe reviewed hundreds of studies on the effects of food irradiation and have concluded that it is safe for consumption.[86] Not only is it safe, but it also helps preserve more of the nutrients available in food, and irradiation is a lower cost preservation method at almost 50 percent less expensive than other conventional methods.[87]

The other tool we have for preserving food is adding certain chemicals that do the job for us. We mentioned citric acid, a common preservative, in the previous chapter. Would you want acid in your food? Well, as you'll recall, citric acid is simply the scientific name for vitamin C, a naturally occurring chemical in citrus fruits that happens to work wonders as a preservative. I think we can all agree that we aren't going to lose any sleep over a little extra vitamin C.

So, what are we worried about when we talk about chemical food preservatives? Let's look at two of the most popular "boogiemen" of

---

86 "Is Irradiated Food Safe?" *Center for Consumer Research.* 28 June 2017. https://ccr.ucdavis.edu/food-irradiation/irradiated-food-safe
87 Ashraf, Shafia, et al. "Food irradiation: A review." *International Journal of Chemical Studies.* 2019. Vol. 7: 131-136.

chemical preservatives: nitrates and nitrites. Both are legal preservatives commonly added to processed meats like bacon, ham, hot dogs, and salami, and even some cheeses. Nitrates and nitrites have gotten a bad name because they were once thought to be carcinogenic (a finding, as we'll discover, that has largely been negated by new studies that show the benefits of these chemicals).[88] This belief led people to avoid foods that had been processed with nitrates and nitrites.

However, as we have seen in other examples, the early findings were issued out of an abundance of caution when these substances were first being studied, and they were based on a misunderstanding of the exact mechanisms by which nitrates and nitrites work in the body.[89] But it is hard to get people to unlearn things they have been taught, and you can still probably find articles online stoking fear about added nitrates (and trying to sell you a pricey "all natural" alternative).

More research has been conducted in the years since those early studies, and a substantial body of evidence now shows the opposite to be true. In 2010, the International Agency for Research on Cancer (IARC) concluded that there is no evidence that nitrates are carcinogenic, and some studies have even shown that nitrates could *reduce* the risk of gastric cancers.[90]

Nitrates and nitrites can actually improve our digestive system by protecting us from certain bacteria. They can lower blood pressure, reduce the risk of cardiovascular disease, and improve recovery time in athletes.[91] There is also a growing body of evidence to suggest that nitric oxide may

88 Ma L, et al. "Nitrate and Nitrite in Health and Disease." *Aging Dis.* 1 Oct 2018; 9(5): 938-945. doi: 10.14336/AD.2017.1207.

89 Corliss, J. "Nitrates in food and medicine: What's the story?" *Harvard Health Publishing.* Harvard Medical School. 1 February 2022. https://www.health.harvard.edu/heart-health/nitrates-in-food-and-medicine-whats-the-story. As it turned out, the ill effects thought to be caused by nitrates/ nitrites were more likely due to overconsumption and thus the intake of high levels of saturated fats and sodium found in processed meats.

90 Ma L, et al. "Nitrate and Nitrite in Health and Disease." *Aging Dis.* 1 Oct 2018; 9(5): 938-945. doi: 10.14336/AD.2017.1207.

91 "Nitrates and Nitrites: What Are They and What Foods Have Them?" *Cleveland Clinic.* 16 November 2023. https://health.clevelandclinic.org/what-are-nitrates

help improve outcomes for conditions like inflammation, diabetes, and dementia.[92] Indeed, because of their positive effect on blood flow, nitrates are commonly prescribed to treat angina.[93]

One of the great mysteries of the nitrate/nitrite debate is that most of our intake of dietary nitrates and nitrites comes from—you'll never guess—leafy greens![94] Yes, that's right. Remember our early discussions about fertilizers and how important nitrates are to plants? Nitrates are the building blocks of plant protein, so it's no wonder that when we eat plenty of vegetables, especially leafy greens, carrots, celery, and beetroot, we are also eating the nitrogen that the plants absorb from the soil. And you'll be hard pressed to find any doctor who suggests you eat *fewer* vegetables as part of a healthy diet.

Indeed, according to a professor of public health at Harvard, "While natural nitrates might sound healthier, that's not the case—your body can't tell the difference."[95] This phenomenon is true of many common preservatives, including sorbic acid (found in berries), benzoic acid (from fruits and spices), and sulfites (found in black tea, peanuts, and eggs). We use synthetic versions of naturally occurring chemicals because it is more efficient and economical for producing the large quantities we need to meet the food supply needs of billions of people.

While we're on the subject, I want to reiterate that "natural" does not always mean better or safer. For example, betanin is a common "all natural" preservative and food coloring derived from beetroot. However, if

---

92 Ibid.

93 Lee PM, Gerriets V. "Nitrates." *StatPearls*. 2024. https://www.ncbi.nlm.nih.gov/books/NBK545149/

94 Dowden, A. "Usually associated with processed meats, nitrates and nitrites are potentially cancer-causing compounds. But what are they, really – and are they always detrimental?" *BBC*. 13 March 2019. https://www.bbc.com/future/article/20190311-what-are-nitrates-in-food-side-effects. According to the report, only 5% of dietary nitrites came from processed meats, and more than 80% came from vegetables.

95 Corliss, J. "Nitrates in food and medicine: What's the story?" *Harvard Health Publishing*. Harvard Medical School. 1 February 2022. https://www.health.harvard.edu/heart-health/nitrates-in-food-and-medicine-whats-the-story

you are allergic to beetroot, betanin can cause a severe allergic reaction.[96] And yet no one calls for us to abandon all natural preservatives as they do with synthetic preservatives like sulfites, which can similarly affect people with asthma and, in rare circumstances, bring on an asthma attack.[97]

If you have a specific dietary requirement or medical condition, of course it is important to be aware of what's in your food, but for most of us, there is no need for concern. Alarmism about preservatives likely stems from our biases or from perceptions based on media misinformation about health and nutrition.

Now, despite all the evidence, you may still be thinking, *well, it doesn't hurt to be cautious. I would rather avoid foods with added nitrates and nitrites on the off chance they could be harmful*, and I understand that impulse. Scientists recognize that they cannot prove that anything is 100 percent safe, but what they can do is develop testing scenarios to ensure that food is safe when we consume it within normal parameters (i.e., not eating 800 pounds of mushrooms in a day).

Of course, there are always outliers or exceptions, but if 95 percent of peer-reviewed studies confirm that preservatives are safe, then the most logical conclusion is that foods with added preservatives are, indeed, safe.

It's important to understand this distinction between normal and abnormal consumption because I think this is where people can easily get lost. We want things to be black and white. We want to be able to categorize things in such a way that we can easily understand them. We want to look at something and determine whether it's good or bad, safe or harmful, and we run into problems when the reality is more complex, when something can be good in small quantities and harmful in large amounts. So, you might think, I'd rather not take the chance. I'll pay

---

96 "Understanding Food Preservatives: What Are the Health Risks?" *Canadian Institute of Food Safety*. 16 November 2021. https://blog.foodsafety.ca/food-preservatives-what-are-health-risks

97 Ibid.

the extra couple dollars, buy the preservative-free bacon, and go on my merry way.

But here's the thing: nitrates and nitrites—and preservatives in general—are necessary because the alternatives are botulism, salmonella, and E. coli. High-risk food items such as meat, seafood, dairy, and cheese are breeding grounds for dangerous microbes because of their high moisture content.

It only takes about 200 parts per million of nitrates to kill the bacteria that causes botulism … or you can skip the nitrates and risk getting food poisoning. So, you have a choice. Do you want to risk getting botulism, which will likely kill you, or do you want to take the extremely low chance that—contrary to the science—years from now, nitrates may cause some carcinogenic reaction in your body at those low levels? I know which option I would pick every time.

One part per million is one inch in sixteen miles. Picture that, and then consider that nitrate levels in most meats must be less than 200 parts per million—and 200 parts per million of nitrates has been evaluated and deemed safe for human ingestion. For context, that is like the length of a stick of gum compared to the distance from New York City to Newark, New Jersey.

I hate to beat this horse to death, but food additives and chemicals used in food production are always assessed for their safety based on the levels you can expect to be exposed to. Food additives are not allowed to be in put in food unless they are deemed to be safe at various levels, and they generally test at levels that greatly exceed normal consumption. Moreover, preservatives are then only used at the lowest effective levels, which are consistently reviewed to keep up with new research.[98]

This remains true whether we are talking about ascorbic acid, nitrates and nitrites, sulfites, sorbates, calcium disodium EDTA, butylated hydroxyanisole (BHA), butylated hydroxytoluene (BHT), or any other

---

98 "Food additives." *Food Standards Agency.* 12 April 2024. https://www.food.gov.uk/safety-hygiene/food-additives

food additive used as a preservative. We use various preservatives to combat different harmful outcomes—like oxidation or spoilage from microbial growth.

Another reason we cannot rely solely on traditional methods like canning and smoking is that traditional methods are often time-consuming. It can take months to properly smoke, salt, or ferment foods, not to mention that they are not as effective as modern methods of preventing bacterial growth and, therefore, are less effective at stopping potential disease outbreaks.

Traditional processes also change the flavor and freshness of food. Consumers increasingly want fresh produce, meat, and other goods because of the higher nutritional content. Traditional methods tend to involve introducing high salt and sugar content, but modern methods deliver where traditional methods fall short. They remove the risk of microbial rot without affecting the food's nutritional content, taste, flavor, smell, or texture.[99]

Lastly, preservatives are critical in the battle against food waste. In the US, 30 to 40 percent of the food supply is never eaten. Globally, nearly one third of the food supply is lost or wasted. That is roughly 1.3 billion tons of food per year.[100]

Because of the impact of food waste on the environment and our energy resources, the EPA has set a goal of reducing food waste by 50 percent by 2030.[101] Preservatives and modern preservation methods are key to achieving this goal, conserving energy and other critical resources, reducing our impact on the environment, and increasing global food security.

---

99 Kumar, Anil. "Food Preservation: Traditional and Modern Techniques." *Acta Scientific Nutritional Health.* 2019. Vol. 3: 45-49. 10.31080/ASNH.2019.03.0529.

100 Gustavsson, J., et al. "Global Food Losses and Food Waste." *Food and Agriculture Organization of the United Nations.* 2011. http://www.fao.org/docrep/014/mb060e/mb060e00.pdf

101 "United States 2030 Food Loss and Waste Reduction Goal." *Environmental Protection Agency.* https://www.epa.gov/sustainable-management-food/united-states-2030-food-loss-and-waste-reduction-goal

If we were to get rid of all preservatives in food, even more food would go to waste before it reached the supermarket. And if it did survive the journey, it would have a significantly shorter shelf life—at the store and at home. Food spoils and loses palatability quickly without the modern miracle of preservatives.

Today, our food is often grown quite far away from where we live, which means it takes time to reach us, time that, without preservatives, would lead to rot and spoilage. So, unless everyone wants to go back to growing all their own vegetables and raising and slaughtering their own animals, then we need our food to be able to last on the journey from farm to table.

For this same reason, preservatives make it possible for us to enjoy a variety of foods that cannot readily grow in our region or climate or at particular times of the year. The next time you enjoy a little wine and cheese night or replicate a dish from your vacation abroad, remember that most of the foods on your table—wine, cheese, charcuterie, or exotic fruits, vegetables, and spices—are only accessible because of preservatives.

So, what do we know about preservatives? Preservatives are yet another critically necessary tool in our fight against Mother Nature. While it is tempting to think that "all natural" is the best option, we must remember that all natural is botulism. All natural is E. coli and salmonella. All natural means food waste and food insecurity.

We are very lucky in North America to be able to drive to the grocery store and know the shelves will be full of fresh, affordable, nutritious food that poses no risk to our health and safety, thanks to a complex system of experts and institutions that evaluate food safety and enforce food safety systems. Microscopic germs are everywhere, and some of them can cause serious illness.

Preservatives are needed to ensure that high-risk products are safe for consumption. And *preservatives* is really the right word to describe them, because what it is really doing is preserving the healthfulness of the food. That we have found a reliable, sustainable way to achieve that outcome is a sign of real progress.

Chapter 9

ARE FERTILIZERS AND PESTICIDES KILLING THE SOIL?

News media will have you believe that conventional fertilizers and pesticides reduce the diversity and quantity of life within soil. As a farmer, I can assure you that is just not true! It's hard enough to farm because so many of the variables you depend on for a healthy crop are out of your control, like weather and crop prices, to name just two. The fact that we do our jobs to the best of our abilities and then have the tools that make our work possible condemned by people who really don't understand the complexity of farming makes things much harder for us. In this chapter, I'll put the case for using fertilizers and pesticides in lay person's terms and demonstrate that such substances don't harm fields, the environment, or consumers.

First, where did fears about fertilizers and pesticides originate? And why have such fears become so mainstream? The main thrust of the argument against fertilizers and pesticides (also called inputs) seems to be that they are "not natural" and, therefore, harmful. Anything synthetic is automatically suspect. This belief rose in parallel with the rise in popularity of organic food, especially among the elite who had greater access and deeper pockets to afford high cost items. The fact that organic food producers decided not to allow synthetic inputs in organic food production gave consumers a reason to doubt the safety of conventional growing methods—even though the choice is largely arbitrary.

What makes it arbitrary? Let's take nitrogen-based fertilizers as an example. If you studied chemistry in high school (or if you were paying attention to the earlier chapters of this book), then you will know that

79

nitrogen is nitrogen—plain and simple. When it comes to the arrangement of atoms in a nitrogen molecule, there is no difference between synthetic and naturally occurring nitrogen. They are identical—whether the nitrogen is found in manmade fertilizer or cow manure or made by lightning. A pound of nitrogen is a pound of nitrogen. So, while organic farmers may use nitrogen fertilizers in the form of cow manure, commercial farmers simply use a synthetic version of the same thing.

The real differences between synthetic and natural fertilizers are the abundance, accessibility, and affordability of synthetic ones. We can produce enough synthetic fertilizers to meet the global food demand, but there simply is not enough naturally occurring fixed nitrogen—the kind of nitrogen available to plants—to grow that much food.

Fritz Haber and Carl Bosch, who devised the process for making synthetic nitrogen, won the Nobel Prize in Chemistry for their discovery, without which almost half the world's population would not be alive today.[102] Let me repeat that: half of the people in the world would not be alive today without the invention and continued use of synthetic fertilizers. The Haber Bosch method is considered one of the greatest inventions of the 20th century—if not *the* greatest.

If we were to farm using only techniques and technologies available before the invention of the Haber Bosch process, it's estimated that we would only be able to feed about four billion people in an era when we are rapidly hurtling toward a population of nine billion.[103] It is literally a life-saving and life-giving technology.

Why is the Haber Bosch process so critical to our survival? There is a limited amount of naturally occurring nitrogen available at any given time (specifically the type of nitrogen that can be used to feed plants as opposed to ambient nitrogen in the atmosphere). In the case of fixed nitrogen produced by lightning, Mother Nature produces about 10 to

---

102 Harford, Tom. "How fertiliser helped feed the world." *BBC News.* 2 January 2017. https://www.bbc.co.uk/news/business-38305504
103 Ibid.

12 pounds per acre per year on average, depending on how many lightning storms there are, and, of course, some regions don't get any lightning storms whatsoever, which means no fixed nitrogen. But even the high end of what can be produced by lightning doesn't scratch the surface of how much you need to grow a high-yield crop.

For comparison's sake, I recently used 110 pounds of nitrogen—made by the Haber Bosch process—per acre to grow 120 bushels of high-protein wheat per acre. Without it, my crop would have produced 20 or 30 bushels, if I was lucky, and it would have a degraded nutritional content. I don't need to tell you that 120 bushels feeds more people than 20 bushels, and because of the principles of supply and demand, higher yields mean more affordable food for consumers.

The same is true of all fertilizers necessary for plant growth. There are not enough naturally occurring sources of these vital chemicals—chemicals that plants *need* to grow—to produce enough food to feed everyone.

Phosphorous is another key element in plant growth, and there are naturally occurring sources, like rock phosphate. Rock phosphate can be mined in relative abundance from places like Idaho, Florida, and Morocco (the number one global producer), and it contains high levels of phosphate minerals, specifically phosphorous pentoxide (P_2O_5). But even after mining rock phosphate, the rocks only contain on average 4 to 20 percent phosphorous pentoxide and otherwise, the phosphorus in rock phosphate is in a form that plants cannot access.

So, one of two things can happen next. Organic growers could embed whole phosphate rocks or spread ground phosphate rocks onto their farm and allow nature to take its course. This is a very slow, inefficient course whereby with rainfall and time the phosphorous will incrementally leach out from the rock and into the soil for the plant to take up. This is an uncommon practice outside of organic growing because of the low availability of phosphorous in the natural material, high transportation costs, and minimal benefit to crops.[104]

Think of this method like going to the junkyard, grinding up a car door, and spreading the car door dust on your land because there is, among other things, zinc in car doors. Plants need zinc, but that is not a good way to get it to them. Like a car door, rock phosphate in its natural form also includes many other things besides phosphate; it contains a whole host of impurities like lead, cadmium, silica, copper, aluminum, iron oxides, and other heavy metals, which also leach into the soil and get taken up by plants—plants that consumers pay four to five times more for.[105] Because of their highly variable chemical constituents, phosphate rocks can release elements into the soil that may be harmful.[106]

Option number two is that rock phosphate can be used to produce phosphoric acid, which is further processed into orthophosphoric acid, the phosphate form that can be taken up and used by plants. The process also gets rid of all those impurities mentioned above. In fact, the resulting phosphorous fertilizer must be tested to strictly ensure that lead and any potentially harmful elements are below the allowable, safe levels. One of my fertilizer suppliers voluntarily shut down their phosphate mine in Ontario when they discovered the rock phosphate was contaminated with lead that they could not remove, which I think demonstrates a real commitment to ethical corporate stewardship.

As a bonus, instead of unwanted heavy metal stowaways in your fertilizer, manufacturers can add in other good, necessary chemicals for plant growth like ammonia, (nitrogen), and potash to synthetic versions.

---

104 "Understanding phosphorous fertilizers." *University of Minnesota Extension*. 2018. https://extension.umn.edu/phosphorus-and-potassium/understanding-phosphorus-fertilizers#process-619211

105 Zapata, F.; Roy, R.N. "Chapter 1 - Introduction: Phosphorus in the soil-plant system." *Use of Phosphate Rocks for Sustainable Agriculture*. 2004 Rome: Food and Agriculture Organization. ISBN 92-5-105030-9.

106 Ibid.

So, is one method better than the other? Absolutely if we're talking about how accessible the phosphorus is to the plant. To put it in context, rock phosphate can have as much as 34 percent total phosphorous content. On average, only three percent of the total phosphorous is phosphorous pentoxide (P_2O_5), which plants can access, but 0 percent—*zero*—is water soluble, meaning that a plant cannot actually take it up and use it.[107]

Let's compare these numbers to some common commercial phosphorous fertilizers. Concentrated superphosphate (CSP) is 45 percent phosphorus, all of which is in the form of phosphorous pentoxide (P_2O_5). Of that, 82 percent is water soluble and, therefore, *guaranteed* to be accessible to plants.[108] Even better is a fertilizer like diammonium phosphate (DAP), which is 47 percent phosphorous and 46 percent phosphorous pentoxide (P_2O_5), of which 100 percent is water soluble and accessible to plants—plus, it has an added 10 percent nitrogen content![109]

As an aside, this should give you a sneak peek into how fine-tuned fertilizer application is nowadays. The label and instructions on any fertilizer will tell you exactly how much of each nutrient is in the fertilizer, how much the plant can access, and how much it needs to grow, and that informs how much we use to feed our crops. It's part of the 4Rs: the right fertilizer source, at the right rate, at the right time, and in the right place.

Another benefit of synthetic fertilizers over natural or organic fertilizers is that they replace the nitrogen that plants extract from the soil. If we don't replace that nitrogen, then the soil and organic matter base in the soil become degraded. By this point in the book, you may have realized just how vital it is that we protect our soil, especially the first six inches of topsoil. This figure varies in certain regions, but generally speaking, those six inches are the most critical.

---

107 "Understanding phosphorous fertilizers." *University of Minnesota Extension.* 2018. https://extension.umn.edu/phosphorus-and-potassium/understanding-phosphorus-fertilizers#process-619211
108 Ibid.
109 Ibid.

Topsoil is where organic matter like bacteria, fungi, and other organisms, which maintain soil quality, are most active and plentiful. It is where most root structures are anchored and where plants get their nutrients. Topsoil also retains moisture, especially during droughts and dry spells, and acts as a buffer against pollutants, thus, protecting groundwater sources.

Synthetic fertilizers are more reliable and more effective at replenishing nutrients in the soil. If we do not replace the nitrogen and other necessary elements in the soil, then plants will struggle to grow … and eventually, *nothing* will grow. Those six inches of topsoil are the most critical factor in the environment, the food production system, and the lives of human beings. Deplete that resource, and we are done.

It isn't just the soil that needs these nutrients. Synthetic fertilizers are key to reducing micronutrient deficiencies or "hidden hunger" in people, too. Fertilizers help fortify staple food crops with micronutrients and alleviate deficiencies in important nutrients like iron, zinc, selenium, and iodine in communities around the world.[110] This strategy, also called biofortification, is considered to be the most sustainable intervention for addressing hidden hunger—more so than dietary diversification and vitamin supplements.[111]

This leads us to the next stage of the nutrient cycle. So far, we've seen how fertilizers add essential nutrients to the soil, those nutrients get taken up by plants, and then, are absorbed by people when we eat those plants. So, what happens next? The nutrients end up in the sewage system. There is a facility near us in Saskatoon that makes phosphate fertilizer from the city's sewage. They sell it to us farmers, who buy it and spread it back on our land. It is more expensive, but we are actually recycling the same phosphate back onto the land.

---

110 Ofori KF, et al. "Improving nutrition through biofortification: A systematic review." *Frontiers in Nutrition.* 9 Dec 2022. 9:1043655. doi: 10.3389/fnut.2022.1043655.

111 Wakeel A, Labuschagne MT. "Editorial: crop biofortification for food security in developing countries." *Front Sustainable Food Systems.* 2021. 5:756296. 10.3389/fsufs.2021.756296

Many new technologies like this are being developed to improve current systems. One criticism of fertilizers and pesticides is that their production can be energy-intensive, but we have to think of this in terms of the energy cycle. Yes, it takes energy to produce fertilizers and pesticides, but they help produce more food, which then gives us the energy we need to live, quite literally. So, we can't simply do away with the inputs that help us grow food. What we can do is improve how we make them.

Scientists, engineers, and innovators are working on just that. One way is by reducing the temperature and pressure currently required for the Haber Bosch process. According to scientists at the Massachusetts Institute of Technology (MIT), "These changes would make it easier to run fertilizer plants entirely on renewable energy or other climate-friendly sources. They would also allow ammonia to be made in smaller factories, making fertilizer more accessible to farmers in developing nations."[112]

Scientists have even developed what is called "green" ammonia or zero carbon ammonia (ammonia meaning nitrogen). According to the Director of United Nations Environment Programme Ecosystems Division, "The sustainable use of nitrogen offers a triple win—for the economy, for human health, and for the environment."[113]

Fertilizers give us the ability to grow healthy, high-yielding crops. The next major hindrance to crop production (or *yield*) is weeds. This is where pesticides come in. You may be thinking, *doesn't "pesticide" mean it only controls "pests" like insects?* But *pesticides*, or "plant protection products" (PPP), is a general term for inputs used to control pests, weeds, and diseases, and they include insecticides, fungicides, herbicides, molluscicides, and plant growth regulators.

112 "Fertilizers Are Necessary." *The Climate Reality Project.* 11 February 2022. https://www. climaterealityproject.org/blog/fertilizers-are-necessary-and-only-evil-we-let-them-be

113 Pradhan, M. "Fertilizers: Challenges and Solutions." *United Nations Environment Programme.* 2024. https://www.unep.org/news-and-stories/story/fertilizers-challenges-and-solutions

Back to the problem of weeds. If you have a garden, you are probably in a constant battle against weeds. What happens to your beautiful tomatoes, carrots, and herbs if the garden is full of weeds? Your plants don't grow because the weeds choke them out. That is because plants have to compete with weeds for nutrients and water in the soil. The weeds steal the water and nutrients, and there goes your yield. You can fertilize all you want, but if you don't control the weeds, the fertilizer will have a very limited impact.

So, how do farmers control weeds? Historically, the only way to control weeds was with a cultivator shovel or hoe. We had to till the soil to kill the weeds. Even as recently as 30 or 40 years ago, herbicides would kill certain weeds but not all of them, so mechanical cultivation was the best tool for getting rid of all weeds.

But what about the topsoil? Doesn't mechanical tillage disrupt the all-important top six inches? Yes. Yes, it does.

Enter glyphosate. First introduced in the 1970s, glyphosate works hand-in-hand with GMO crops that have been engineered to be resistant to the pesticide in glyphosate. Farmers can apply glyphosate to their crops, the glyphosate kills all the weeds, and crops thrive. We don't need multiple pesticides, just the one. Most significantly, we no longer need to till (my farm is zero tillage), which protects the topsoil and all the beneficial microorganisms and organic matter in it.

One of the most pernicious myths about glyphosate is that farmers were spraying too much , the runoff was carried into rivers and eventually, the ocean, and that people who swam in waterways exposed to glyphosate were found to have measurable levels of the chemical in their urine. Now, it is true that glyphosate has been found in rivers and surface water around the globe; however, the cause, we now know, is much more likely to be household use of laundry detergent—not glyphosate applied to crops.[114]

---

114 M. Schwientek, et al. "Glyphosate contamination in European rivers not from herbicide application?" *Water Research*. 2024; 263(122140). https://doi.org/10.1016/j.watres.2024.122140.

Researchers were curious as to why there was no reduction in glyphosate levels in European rivers following the adoption of measures to reduce glyphosate use on nearby farms. They also expected to see increased concentrations of glyphosate in spring and autumn when farmers apply the herbicide and following heavy rainfall, but that was not the case. That meant the primary source of glyphosate in rivers had to come from somewhere other than farms.

Through a comprehensive metanalysis of glyphosate concentration patterns from roughly 100 locations in rivers in Germany, France, Italy, Sweden, Luxembourg, the UK, the Netherlands, and the US, researchers found that aminomethylphosphonic acid (AMPA), a chemical commonly found in laundry detergent and other household products, was being converted into glyphosate in wastewater treatment plants. Laboratory tests confirmed the hypothesis.

The finding also stands to reason given glyphosate's nature. In agricultural applications, it tends not to "transport" far because it binds to organic matter in the soil and gets consumed by microorganisms. In other words, it largely stays where it has been sprayed.

Today, glyphosate is one of the most widely used pesticides on the market because it is one of the best tools we have. It is low cost, one of the safest products we work with, and it allows us to make money, which is an essential part of farming sustainability. Farmers need to make money if we want to stay in business. On average, farmers in the US only make $0.14 out of every dollar that consumers spend on food.[115]

But if glyphosate kills weeds, can't it harm the environment or worse, us? Clearly, the answer is no, but let's look at how we know this. Cornell University developed the environmental impact quotient (EIQ), a formula that provides growers with data regarding the environmental and health impacts of over 500 pesticide options so they can make better-informed

---

115 "The Farmer's Share." *National Farmers Union*. 2024. https://nfu.org/farmers-share/

decisions regarding their pesticide selection. The quotient assigns values to each of the individual effects of a pesticide to measure the impact.

The EIQ of glyphosate is 15.33, which is considered low.[116] If you recall our discussion of spinosad in Chapter 2, a pesticide approved for use in organic farming, its EIQ is 14.38, less than one point lower than glyphosate on environmental impact.[117] Soap, which we all use daily, scores around 19.5. I recently heard a nonfarmer state that farmers are killing earthworms with glyphosate and chemical fertilizers. The book of organic composting says that you require six to eight earthworms per cubic foot of soil for it to be healthy.

On my farm, I have six to eight earthworms per cubic foot of soil on every acre! That's about 305,000 earthworms per acre, and I have been using glyphosate and chemical fertilizer for almost 30 years. By the way, did you know that earthworms are not naturally occurring beasts in North America? At least, wherever there was a glacier! That's right. Earthworms are an invasive species to soils that were covered with glaciers in North America. The earthworms got here "unnaturally" in soil that was used to ballast the first ships that arrived on our shores, and they spread from there. We spread them! I can remember as a child that every time we went to visit my grandmother, we were required to go dig earthworms from her hearty garden and transfer them to ours to make better soil.

Naturalists reading this might want me to figure out how to kill them all now because they are not "natural!" Sorry, not a chance. I like them for my soil health, unnatural or not!

As time progresses, our use of pesticides continues to become more sophisticated and precise. And as we saw with fertilizer, new technologies are helping us achieve the goal of using fewer pesticides.

How much crop protection pesticide do I actually use on my farm? I heard a speaker at a conference say, "Farmers are pouring chemicals onto

---

116 "EIQ Pesticide Values." *Cornell University College of Agriculture and Life Sciences.* 2024. https://cornell.app.box.com/v/eiq-pesticide-list

117 Ibid.

the land and food," giving one the impression that these compounds are literally running off and soaking everything around. Let me put things into perspective. Let's add up all the crop protection active ingredient chemicals I use in one full year of growing canola, for example, on one acre of land.

Take a 15 mm eyedropper that you might have in your medicine cabinet. Try to make just one small droplet of water with that. Now, divide that droplet of water into three and try to spread that small amount onto one 8.5" x 11" piece of paper (about a square foot). That is about how much total chemical would hit one square foot of plant and soil in a full year! That's hardly pouring it on! And remember, the soil and plants that it hits will immediately start to break it down. Pesticides can absorb into the plants they hit, removing them from the environment and preventing them from becoming a water contaminant.

Pesticides that hit the ground will adsorb soil particles, i.e., the particles will adhere to the surface of the pesticide. Once bound to the soil, it is unlikely they will leach or runoff. The longer they are bound there, the more likely that microbiological degradation will occur, which further lessens the risk of leaching or runoff. All these mechanisms are tested to ensure their fate in the environment is safe to the point that they do not cause unwanted outcomes. Ultimately, they break down or degrade into simpler compounds, which are usually less harmful to the environment.

Remember, pesticides are a huge operational cost for farms, accounting for as much as 50 percent of a farmer's operational budget.[118]

Some are relatively low-tech practical solutions, like the storage system I patented for glyphosate that I call 'The Racketeer.' Agrochemicals are often shipped in plastic containers as concentrates that are then diluted at the farm. Some portion of the chemicals can get stuck inside the plastic containers after most has been poured out. The concentrated remnants are toxic, which creates a problem regarding packaging disposal. I've designed

---

118 Evans Ogden, L. "The clean farming revolution." *BBC.* 2019. https://www.bbc.com/future/bespoke/follow-the-food/the-clean-farming-revolution/

a safe way to handle glyphosate, store it on the farm, and distribute it while eliminating the plastic packaging.

Some innovations are high-tech, like John Deere's "See & Spray" technology, which uses computer vision and machine learning (artificial intelligence) to locate weeds and only apply pesticides directly onto those weeds. This saves farmers money because pesticides are expensive, and it can reduce pesticide use by as much as 77 percent.[119]

Or look at Carbon Robotics' LaserWeeder, which uses artificial intelligence to identify weeds and then kills the weeds with laser beams, all without disturbing the soil or affecting the organic matter in the topsoil. There are no chemicals involved in this process, which again can help reduce costs for farmers.

It is exciting to see new tools that can help us reduce pesticide use, but it is also important to remember that technologies must be scalable and accessible. Thus, pesticides are still the best option for the global market. We can't expect small farmers or people in remote regions to be able to readily transition to these futuristic tools, which have significant up-front costs and require skilled maintenance. And we don't need them to because fertilizers and pesticides are safe and effective, and because low-input and minimum-tillage or zero-tillage agriculture are also recognized as "nature-positive" and regenerative practices.[120]

The bottom line is that fertilizers and pesticides enable us to grow more food on a fixed amount of land, and that means more efficient land use. To paraphrase Mark Twain, the thing about land is that they are not making it anymore. Every time we pave over a square foot of soil to build shopping centers, housing developments, or office blocks we are removing a square foot of food production. Bear in mind that most cities are built

119 Dulaney P, et al. "See & Spray Technology." *Mississippi State University*. 2023. https://extension.msstate.edu/publications/see-spray-technology

120 Pradhan, M. "Fertilizers: Challenges and Solutions." *United Nations Environment Programme*. 2024. https://www.unep.org/news-and-stories/story/fertilizers-challenges-and-solutions

on some of the best topsoil in the world because cities have historically been built along river deltas. Urban expansion and suburban sprawl are as much threats to the environment as anything else.

But farmers know how valuable land is. It is in our interest to protect it and make the best use of it we can. In the US, conscientious stewardship by farmers has driven a 34 percent decline in erosion of cropland since 1982.[121] Moreover, the use of pesticides and fertilizers improves the efficiency of land use—meaning we do not need to clear more land, especially from important ecosystems like the Amazon, to produce enough food to feed everyone. Instead, we can produce higher yielding crops on existing farmland using the technology available to us, and still protect the biodiversity of wild places.

Natural fertilizers simply cannot compete with the effectiveness and reliability of synthetic fertilizers and pesticides to help grow nutrient-dense plants and produce high yields. Despite the obvious advantages of synthetics—and the fact that many are derived from natural sources—they are baselessly demonized because they are produced through an industrial process.

I like to give people the benefit of the doubt. People want to buy food produced without synthetic fertilizers and pesticides in large part because of successful marketing campaigns by Whole Foods, the organic industry, and influencers who stand to profit from selling you expensive alternatives. There is also a kind of elite status associated with foods branded as "organic" or "pesticide free." But the point I want to drive home is that all these signals of elite status—signals that are meant to tell us that our food will not cause harm—have a real cost. A lethal cost.

There is nothing wrong with composting or using cow manure, but they are not scalable options for the global food supply. Synthetic fertilizers are critical to feeding the world. Demanding that farmers abandon fertilizers and pesticides in food production would doom half the world

---

121 "Fast Facts About Agriculture & Food." *Farm Bureau.* 2024. https://www.fb.org/newsroom/fast-facts

to starvation. It is, without exaggeration, an endorsement of letting millions of people starve.

Our choices as consumers can dictate public policy. If enough of us try to convince our representatives not just to regulate fertilizers and pesticides (and they already are very strictly regulated) but also to forbid farmers from using essential inputs like nitrogen and glyphosate, then we will destroy the global food supply and cause mass starvation. It sounds extreme, I know, but those *are* the stakes.

Fertilizers and pesticides are safe. And more importantly, they are hugely beneficial and essential if we want to meet the food supply needs of the global population.

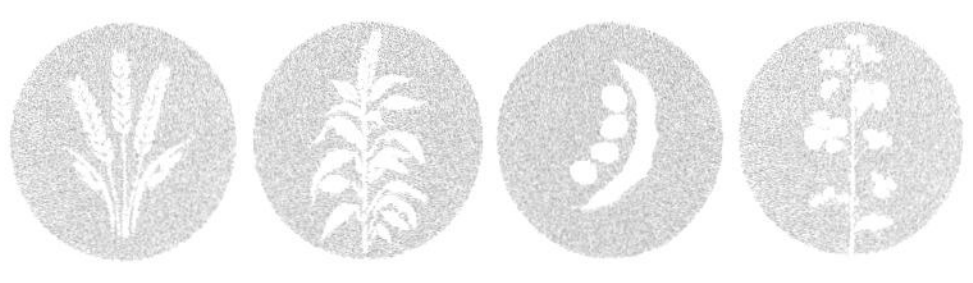

Chapter 10

SIX-STEP PROCESS TO ETHICAL AGRONOMY

We've talked a lot about the myths surrounding farming and food production, and we've taken a largely consumer-focused approach by examining the safeguards implemented by governments, regulatory bodies, and third parties to ensure our food supply is safe. But what about farming from a farmer's perspective? I've shared nuggets of my thoughts and approaches, and now I want to broaden that scope.

How do farmers approach the concept of farming? What are we doing to improve how we grow food? There's far more to it than planting seeds, harvesting crops, and selling them for profit. Farmers make painstaking efforts to consider their impact on the environment, consumers, and the future of farming. In fact, modern methods sequester carbon and boost biodiversity.

In this chapter, we'll explore the six elements of *ethical agronomy*, the practice of treating the soil properly to get optimal results, improve yield, protect the environment, and help farmers to budget.

Agronomy is simply the science of farming. It includes soil science, biology, ecology, chemistry, animal agriculture, genetics, pest management, water science, and even economics—anything and everything to do with the science and technology of agriculture.

Ethical agronomy is a holistic management system that looks in-depth at the six main factors that affect crop production, with the goal of growing crops effectively and profitably while conserving natural resources and protecting the environment. The six factors of ethical agronomy are seedbed preparation, seed quality, seed placement, seed date, crop nutrition, and crop protection. There are also numerous sub-factors within each factor.

By addressing each of the six factors individually, growers can implement best practices based on cutting-edge research and ensure they get the highest yield possible from their land. To this end, I've helped to develop the Farm Performance Assessment (FPA), which assesses and scores farmers on each of the six factors so they know what levers they can pull to improve their yield, profitability, and sustainability while minimizing their environmental impact. It also gives us a little peace of mind knowing we've done everything we can to mitigate risk and ensure the future viability of our farms.

Another key element of ethical agronomy is the link it creates between agronomists and growers, creating channels for agronomists to communicate the newest, most reliable science to growers in a way they understand. It also helps growers make informed, foundational decisions that affect not only their bottom line but also the overall productivity of their farms. Yes, farming requires many advisors and experts to make sound, safe, and profitable management decisions. When I wrote this book, I had trusted experts and editors review it. It is the same with farming; the more input and advice you get before making a decision, the better the odds that your decision will be an informed, sustainable, and positive one for the soil, the food it grows, and the profitability of the system.

What makes six-factor agronomy ethical? I'd like to reiterate a concept we discussed in Chapter 5, "A sustainable agriculture must be economically viable, socially responsible, and ecologically sound … and the three must be in harmony."[122] By taking a whole-farm approach, ethical agronomy gives farmers better control over the factors with the greatest outcome on their sustainability, profitability, and environmental impact.

* * *

---

122 Ikerd, John D. "What Is Sustainable Agriculture?" Sustainable Agriculture Research and Education. https://western.sare.org/about/what-is-sustainable-agriculture/

Let's explore each of the six factors of ethical agronomy, starting with *seedbed preparation.*

Before we get anywhere near to planting our seeds, we need to prepare the seedbed. The seedbed is exactly what it sounds like—it's the environment where the seed is placed. A high-quality seedbed allows the crop to germinate quickly, sprout from the ground, cover the ground, and achieve its genetic potential.

Crop rotation is a big part of the score for seedbed preparation because it impacts soil quality and biodiversity. So, it is important that we look at the crops that a grower grows, in what order they grow them, and over what period and score them on a scale of one to ten (one being "poor" and ten indicating "great").

In western Canada, we generally have about a four- to five-month growing season. Further south, we might find nine-month growing seasons, meaning farmers can "double crop" or grow two rounds of crops per year.

Crop rotation is the practice of growing one crop on a plot of land, and then growing a different crop the next year on the same land. North of Saskatoon, we have a wet, cool climate, so growers often choose to rotate between canola and wheat. A little farther south, where it's drier, farmers might grow durum one year and lentils the next. These represent very simple rotations.

Increasingly, farmers are adding a "pulse crop" to their rotation, such as peas, lentils, or chickpeas. When we do this, we apply a species of bacteria to the seed that creates a symbiotic relationship between it and the plant. The bacteria infect the roots of the crop and pull nitrogen from the atmosphere into the plant. As a result, we don't need to add nitrogen fertilizer to that crop in that year. So now, we grow wheat one year, peas the next, and canola the year after. Or we can add even more complexity with four-year or eight-year rotations of different plant varieties.

With each addition to our crop rotation system, we increase the soil biodiversity. How so? Every plant has a different *microbiome* comprising all the different microorganisms—like bacteria, fungi, and viruses—that live in and around the environment of the plant. So, by planting different crops each year, different microorganisms are introduced into the soil.

This is important for many reasons, not least of which is disease prevention. Root rot is a common disease in the area where I farm, but rich biodiversity in the soil strengthens a crop's immune system. It's like your pet dog, if you have one—a crossbreed like a Labradoodle or a lovable mutt from the pound is going to have greater genetic diversity than a pure-breed dog, and the greater the genetic diversity, the better your dog's chances of avoiding genetic diseases and conditions. The same is true for plants. As we increase the diversity of our crops and the timeframe of crop rotation, the overall health of the soil and the crops improves.

Straw management is another step we can take to prepare a strong seedbed. During harvesting, the combine separates the kernels we collect from the stalk of the plant resulting in "straw" that gets spread out behind the machinery.

The straw is trash, and we love trash in agriculture. It is high-carbon plant material that helps build organic matter in the soil and protects our land from the elements. Our goal is to spread the straw evenly so that at seeding time, the crop can emerge uniformly and quickly. If we do a poor job of spreading the straw, some seeds will sprout tomorrow and some a week from now, resulting in an uneven, ragged crop. Whether a crop grows in unison affects productivity, because it is harder to properly nourish and harvest a crop when individual plants within that crop grow at different times.

The last piece of the puzzle critical to seedbed preparation that we will discuss here is moisture management (there are several more sub-factors in seedbed preparation, but I'm giving you a general impression of what gets considered in the six-factor system). In dry-land agriculture, moisture is determined by rain or snowfall. If the moisture exceeds a crop's requirements, it can kill them. The same is true if there is not enough rain or snowfall. It is very much a Goldilocks problem—we want just the right amount of moisture for our yield targets.

A farm's moisture management strategy depends on its geographical location. Places like Montana, Wyoming, and the Central Great Plains of North America are very dry regions and are, therefore, considered very conducive to good moisture management because they need and use every drop of water to feed the plant. We have tall crop stubble (the straw and chaff left on the ground), which protects the soil and seeds from wind and wind evaporation and keeps the soil cool, which further prevents moisture loss due to evaporation caused by heat from the sun. In other words, we have maximized the amount of water and snow we receive naturally to maintain the ideal moisture content.

In contrast, corn country in Iowa gets an average of 24 to 26 inches of annual precipitation, thus, there are times when crops receive too much moisture. In response, farmers have to expose some of the soil by tilling, which encourages evaporation. They can manage moisture this way, but as we've learned in earlier chapters, the downside to tilling the soil is that it disrupts the top six inches of topsoil and the critical organic matter in it.

To sum up: crop rotation, straw management, and moisture management are critical for the success of seedbed preparation. That is step one of six-factor agronomy.

* * *

The second factor is *seed quality*, meaning the genetic potential within a seed. Farmers today need to know several seed quality parameters to plan for a successful crop.

When a grower is planning a crop, it is important that they understand the germination, vigor, and weight of the seed to ensure they are calculating the proper seeding rate. A proper seeding rate ensures that there aren't too few or too many seeds being placed in the soil, as this ultimately affects crop yield.

* * *

Seed placement is the next factor impacting the success and yield of a crop. Farmers need to decide the depth at which the seed is placed, its proximity to "stubble" or last year's rows of seeds, and its proximity to other seeds. What are the ground conditions best suited for planting? Should the soil be wet, dry, or in between?

Corn is a great example of the importance of seed placement because corn is extremely responsive to changes in the placement. The physical depth from the surface to where the seed is placed can affect yield. We have equipment today that can change that depth in increments of an eighth of an inch.

Most of the nutrients in soil will be within the top three inches of topsoil. The closer the nodal roots of a corn plant are to the surface of the soil, the harder it is for that plant to absorb nutrition because the soil surface dries quickly after rainfall. Most plant nutrients are water soluble, so as the soil dries, there are fewer available nutrients.

But if you can, plant the seed so the nodal roots are deeper in the ground, where the soil stays moist for longer and the plant can soak up more nutrition. Again, the precise "Goldilocks" zone for seed placement will vary from farm to farm.

* * *

The final factor focused on the seed is the *seed date*. When is the optimal time to plant the seed? The day a farmer chooses to put their crop in the ground is absolutely critical to its success. The earlier the better, and the warmer the soil the better.

When land is ready to be seeded, earlier-seeded crops will generally outperform later-seeded crops in terms of yield. Knowing when to seed can, therefore, increase your yield without costing you anything extra. It is a free way to improve your productivity and profitability.

* * *

Moving beyond the seed, we need to fertilize the crop so that the seed—and later the plant—gets proper *nutrition*. As we've discussed in earlier chapters, plants need sufficient nutrients to grow—nitrogen, phosphorus, zinc, and so forth—and a properly nourished plant will pass on its nutritional value to us human beings when we eat it.

The key to ethical agronomy is to match the nutrition to the target yield and apply nutrients in accordance with the four Rs: the right rate, the right form, at the right time, and in the right place. This way, the plant gets exactly what it needs—nothing more and nothing less. We determine what it needs by conducting an annual soil test analysis and tissue test results from previous years. Amazingly, I actually think we feed our crops with more diligence and detail than we feed ourselves!

* * *

Lastly, we need to *protect* our crops from weeds, diseases, and insects that steal resources from the plants or destroy them entirely. This is critical because a farmer can do absolutely everything else right—prepare the seedbed perfectly, buy the best quality seeds, plant them at the right time, and fertilize according to the four Rs—and still have a zero-yield crop if they fail to protect it. Furthermore, crop protection involves both chemical inputs (like glyphosate) and cultural practices we can implement (like zero tillage).

Ideally, we want to have a weed-free crop from the time the seed germinates right up to harvest meaning all the resources invested into the crop are used only by the crop and not by weeds, disease, or insects.

The good news is that our ability to protect crops is improving all the time. For example, this year, I did not need to use any chemical crop protection. I don't like using chemical crop protection because, frankly, it's expensive. Instead, I used a crop that was conventionally bred to control weeds; the plants have a gene that makes them secrete a compound from their roots that suppresses other plants (weeds) from growing.

So, for example, I seeded my rye crop last fall. It spent the winter underground as a living crop under a cover of soil, straw, and snow. Then, as soon as spring arrived and the crop saw sunlight, it started growing, way ahead of any weeds. I did not need chemical crop protection because the crop was protecting itself. Rye actually has allelopathy. It secretes a substance that impairs the growth of other plants, which is a form of natural weed control!

* * *

The goal of ethical agronomy is to find the right balance between all six factors so we get a strong, healthy crop that meets its yield potential. Seedbed preparation, seed quality, seed placement, seed date, crop nutrition, and crop protection must all be optimized to get the maximum yield.

If we make a poor management decision regarding seed placement, it can have a ripple effect on the bottom line, i.e., the farm's overall productivity. If we make multiple poor management decisions across all six factors, it can have a profound effect on profitability and production and our ability to sustain the farm in the long term.

In the FPA, each factor gets scored and the combined scores predict yield based on an algorithm that is correct to within three to five bushels per acre of the actual yield. There is a very narrow margin of error.

For example, if we know the crop production parameters and the algorithm tells a farmer they can get 45 bushels to the acre, then we know the farmer will get between 40 and 50 bushels per acre in terms of real yield. The six-factor score provides something like a checklist for the farmer to follow and gives growers a higher level of confidence in their decision-making.

It is also worth noting that this is the first predictive algorithm developed for and applied to agriculture, which isn't to say that folk in Silicon Valley haven't tried!

It's an incredible tool because, for example, if your initial crop assessment shows your farm producing a yield of 38 bushels but your goal is 50 bushels, then you can go back through your FPA and find variables, like variety, crop rotation, or straw spread, that you can tweak to reach your goal. We are finding better and better ways to farm all the time, and six-factor agronomy is one way of sharing best practices and new technology with other farmers.

I've seen the results first-hand. One of my farmer clients had five extremely dry seasons in the southwest portion of the province where he farms. Four years ago, he did an FPA and scored low on the first factor, seedbed preparation. So, he started the process of trying to build up the "trash" level, i.e., the straw and other waste left on top of the soil at the end of harvest, for the lentil crop. He focused heavily on straw and moisture management and managed to increase his crop yield to about 23 bushels per acre.

Meanwhile, his neighbors, who seeded their lentils at the same time, only yielded about 10 or 15 bushels per acre. Those extra eight bushels per acre make a huge difference to profitability and sustainability.

We make a lot of changes with an eye on the future to stem the tide of climate change. For example, my farm is zero-tillage and direct seed, which sequesters carbons. I sequester about a ton of carbon per acre. We actually take carbon out of the atmosphere with our roots and direct it into the ground, and because we don't disturb the soil, we don't release the carbon back into the air.

Meanwhile, large corporations are paying to buy carbon credits without noticeably changing their behavior or business models.

So, why do we do it? Why spend all this time thinking about how to get 10 or 12 more bushels per acre? It may not seem like a lot; after all, a bushel of wheat only weighs about the same as a Labrador or an eight-year-old kid.

However, one bushel yields enough flour to make 70 loaves of white bread or 90 loaves of whole-wheat bread.[123] When we extrapolate from there, those extra 10 or 12 bushels per acre have the potential to feed an awful lot of people while simultaneously making better use of one of our most limited resources: land.

Let's take Kansas as an example. Kansas is the number one wheat producer in the US. An acre of Kansas wheat produces enough bread to feed nearly 9,000 people for one day.[124] Each year, the state produces enough wheat to bake 36 billion loaves of bread, enough to feed everyone in the world *but only for about two weeks.*[125]

There's only a 60-day supply of wheat in the world.[126] If one wheat-producing area experiences a disaster like a drought or war, it puts an incredible strain on the whole supply. That's how vulnerable we are to mass starvation—there are just two months' worth of wheat sitting in storage silos to make enough bread for people to eat.

And those grain stores are not equally distributed, so some places in the world would feel the effect way more than others.

I think about this all the time. One in ten people in the world still go to bed hungry without enough calories or nutrients to sustain a healthy body. As agronomists, we are trying to maintain optimal yields so there is enough food in the world to feed and nourish *everyone.*

The second vital reason we need ethical agronomy—and why agronomists spend all our time thinking about these seemingly minor details—is to protect our soil and waterways.

---

123 Debes, Julia. "What Does a Bushel of Wheat Mean to Me?" *Kansas Wheat.* 2024. https://kswheat.com/news/what-does-a-bushel-of-wheat-mean-to-me

124 "Quick Facts About Wheat." *National Association of Wheat Growers.* 2024. https://wheatworld.org/facts-about-wheat/

125 Ibid.

126 Some estimates put this closer to seven weeks, or 70 days, depending on grain storage inventory levels. See: Hoffman, Jenna. "Is the World Really Running Out of Wheat?" *AgWeb.* 7 June 2022. https://www.agweb.com/news/crops/wheat/world-really-running-out-wheat

Soil erosion is one of our biggest concerns as agronomists. We are charged with ensuring that the nutrition, crop protection, and dirt stay where we put them. Soil isn't just a factor of production, it is also an environmental asset that quite literally sustains life on Earth, including our own.

What's more, 98 percent of farms in North America are family farms like mine. I'm the fourth generation, so you'd better believe we have a vested interest in leaving the soil in better shape than when we started. This was drilled into me by my father, who had it drilled into him by his father. This means we are always working to protect the environment and the soil so the food we produce from it is as healthy and safe as possible.

The other side of the equation is water. I live near the Saskatchewan Basin, which feeds into Hudson Bay. When water melts off a glacier, it travels down through all the waterways for thousands and thousands of miles. On its way, it passes through agricultural landscapes providing water to communities for drinking and irrigation. Therefore, it needs to be safe.

If we apply proper agronomic techniques then people have healthy, clean water to drink and to water their fields with. If we do our jobs wrong, siltation and erosion can occur, which introduces organisms into the water that are hard to remove and can be harmful to people.

Finally, we think about all this because we need farms to be profitable. We want to help growers stay in business because farmers literally keep us all alive. If a farmer is not making money, they go out of business, and that means they are no longer growing food. That's a loss to everyone.

This really is how we agronomists think. It isn't a conscious choice; we do it naturally. It's part of day-to-day life for us. I spend my days thinking about moisture management, phosphate, soil temperature, and the amount of bacteria in my soil. I think about how to maintain vegetative cover and field shelter belts, reduce tillage, and avoid overgrazing.

And most of all, I think about how to protect those top six inches of soil. It takes between 200 and 400 years to produce *one millimeter* of topsoil.[127] That's the width of your fingernail. If you want an inch of topsoil, that's going to take 1,000 years.[128]

In an ethical agronomic system, we aim to find the right products to generate the right yields and the right profits. We are trying to design the perfect system for each individual grower by considering how to grow crops effectively and profitably while conserving natural resources and protecting the environment.

---

127 Arsenault, Chris. "Only 60 Years of Farming Left If Soil Degradation Continues." *Scientific American*. 5 December 2014. https://www.scientificamerican.com/article/only-60-years-of-farming-left-if-soil-degradation-continues/
128 Ibid.

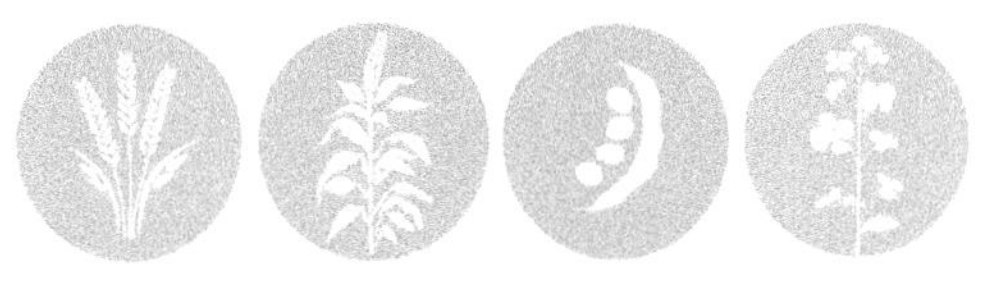

Chapter 11

THE TRUTH ABOUT CANOLA

I met two women at a recent conference, both in their twenties. We were sitting outside on the steps of the convention center during a breakout session. As it was a networking event, I asked about their professions. One woman was from Colorado. She had hair dyed the color of mountains at sunset in deep hues of purple and blue. She had a sponsorship of some kind that sent her traveling around the globe. The other woman operated a supplement company selling something that apparently cures coughs and colds and makes childbirth a pleasure.

They then asked me what I did for a living. I told them I was a farmer from Canada.

"What do you grow?" they asked.

"I grow a lot of things, including lentils, peas, wheat, fava beans, barley, and canola."

Their faces dropped.

"You mean those yellow fields of poison?" the purple-haired one said with a glare. They stopped speaking to me then and there and turned away. I was confused, and frankly a little embarrassed. I didn't know what to do at that moment. The experience made me tremendously sad.

A growing number of people have been led to believe that canola oil is unhealthy, unsafe, or even "toxic" (despite all the evidence to the contrary). Claims against canola oil stem largely from the facts that most canola grown in North America is genetically modified and that canola oil is highly refined to extend its shelf life—and both GMOs and "highly refined" products are often arbitrarily labeled dangerous. Canola critics also assert that because it is high in omega-6 fatty acids, canola oil can increase inflammation.

So, what is the real story here? First, we know that blanket arguments against GMO foods are categorically incorrect and unsubstantiated, as we discussed in Chapter 3. Still, without relitigating that entire discussion, it is worth exploring a little further exactly what kinds of genetic modifications have been made to canola so we have a better understanding.

Canola was originally developed as an edible version of rapeseed because pure rapeseed oil has high levels of erucic acid and glucosinolates, which can be toxic. Canola, however, reduces the levels of these compounds so the oil is safe for consumption. The new plant was developed in Canada, hence the name canola, which is a portmanteau of *Canada,* and the Latin-derived suffix *ola,* which means "oil." Canola is in the same plant family as broccoli, kale, and cabbage—all vegetables we consider "healthy" foods.

While most canola grown in North America is GMO, it is largely the kind of genetic modification we have talked about at length in this book, i.e., a single gene edited to make the plant resistant to certain herbicides so the plant can defend itself against weeds. And we know from previous discussions that GMOs like canola are extensively researched, tested, and verified to be safe.

As it stands, canola is one of the healthier cooking oils on the market. It is low in saturated fats and high in mono- and poly-unsaturated fats. Here is where we run into one of the first problems facing canola oil. Health and diet experts can never seem to agree on what amount or types of fats are good for us, if any.

I can still remember the "low fat" diet crazes of the 1980s and '90s, when everyone ate grapefruit and low-fat cottage cheese and we were all miserable. Companies like Slim Fast, Weight Watchers, Healthy Choice, and Snackwells all piled on to cash in on the diet trend. Pretzels were good (no fat), but nuts were off-limits (high fat).

However, the trend was based on an incomplete understanding of diet and disease. Scientists at the time had some evidence that foods high in saturated fats could lead to higher cholesterol, and higher cholesterol seemed to mean a higher risk of heart disease and heart attacks.[129] But the conclusion they arrived at—that all fat was bad fat—was based on very little data and even less understanding of the complexities of health. Decades later, we found we were not any healthier for cutting out fat—probably because we cut out the *healthy* fats along with the rest.[130]

Thankfully, the science caught up, and now we know that we need some fat in our diet. In fact, certain types of fat are good for us. Several essential vitamins and minerals are only fat-soluble, meaning we need fat to absorb those nutrients from food.[131]

Fats are also a good source of energy, especially if you are active. One gram of fat—*any fat*—provides nine calories of energy, which is more than double the four calories you get from carbohydrates or proteins.[132]

So, let's look at the breakdown of canola oil. According to the USDA, one tablespoon of canola oil contains 14 grams of fat total: 1 gram of saturated fat, 9 grams of monounsaturated (omega-9) fats, 2.7 grams of omega-6 fatty acids, and 1.3 grams of omega-3 fatty acids.[133]

What does that mean for us? Let's start with saturated fats, which are solid at room temperature and common in foods like red meat, cheese, and coconut oil. Canola oil has less saturated fat than most cooking oils, including olive oil, vegetable oil, avocado oil, coconut oil, sesame oil, and butter.[134] That's a positive mark for canola since saturated fats can raise

129 Aubrey, Allison. "Why We Got Fatter During the Fat-Free Food Boom." *NPR, Morning Edition.* 28 March 2014. https://www.npr.org/sections/thesalt/2014/03/28/295332576/why-we-got-fatter-during-the-fat-free-food-boom

130 "The truth about fats: the good, the bad, and the in-between." *Harvard Health Publishing, Harvard Medical School.* 12 April 2022. https://www.health.harvard.edu/staying-healthy/the-truth-about-fats-bad-and-good

131 "Fat: The Facts." *National Health Service.* 14 April 2023. https://www.nhs.uk/live-well/eat-well/food-types/different-fats-nutrition/

132 Ibid.

133 "Oil, canola." *U.S. Department of Agriculture FoodData Central.* April 2018. https://fdc.nal.usda.gov/fdc-app.html#/food-details/172336/nutrients

134 *U.S. Department of Agriculture FoodData Central.* https://fdc.nal.usda.gov/

our "bad" cholesterol or "LDL" cholesterol, increasing the risk of heart disease.[135]

Most fats in canola oil are monounsaturated, which are in a liquid state at room temperature. Monounsaturated fats are considered "heart-healthy fats" because they help lower our LDL cholesterol and reduce the risk of heart disease.[136] In fact, canola oil is one of the top contenders in this regard, boasting some of the highest levels of monounsaturated fats among popular cooking oils. It is also rich in phytosterols, which reduce the absorption of cholesterol into the body.[137] So far, so good.

What critics claim is the "toxic" part of canola oil is the omega-6 fatty acid content. High levels of omega-6 fatty acid intake have been associated with increased inflammation. But this is where critics of canola oil seem to stop reading. The science tells us that the impact of omega-6 fatty acids on our bodies depends on the ratio of omega-6 to omega-3 fatty acids.

The golden ratio for omega-6 and omega-3 fatty acids is 4:1.[138] Our diets have struck this balance since the Paleolithic Era, and until about 100 years ago, that was still the average intake ratio for most people.[139] The problem today is that the typical Western diet provides an omega-6 to omega-3 ratio of approximately 20:1.[140]

You might be thinking, "Gotcha!" But here's the thing—look back at the nutritional values of canola oil. It has a 2:1 ratio of omega-6 to omega-3 fatty acids. That is significantly better than what is considered the healthy

---

135 "LDL and HDL cholesterol and triglycerides." *Centers for Disease Control and Prevention.* 15 May 2024. https://www.cdc.gov/cholesterol/about/ldl-and-hdl-cholesterol-and-triglycerides. html?CDC_AAref_Val=https://www.cdc.gov/cholesterol/ldl_hdl.htm

136 Ibid.

137 Crosby, Guy. "Ask the Expert: Concerns about canola oil." *The Nutrition Source, Harvard T.H. Chan School of Public Health.* 13 April 2015. https://nutritionsource.hsph.harvard. edu/2015/04/13/ask-the-expert-concerns-about-canola-oil/

138 DiNicolantonio JJ, O'Keefe J. "The Importance of Maintaining a Low Omega-6/ Omega-3 Ratio for Reducing the Risk of Autoimmune Diseases, Asthma, and Allergies." *Missouri Medicine.* 2021 Sep-Oct;118(5):453-459. https://www.ncbi.nlm.nih.gov/pmc/articles/ PMC8504498/

139 Ibid.

140 Ibid.

standard. So, the argument that canola oil (consumed at proper daily levels of oil in a diet) causes inflammation simply does not hold water.

The closest that anti-seed oil proponents come to scientific observation seems to be that people who eat less seed oil tend to be healthier. This could be because such people are eating less oil in general. "People who control their diet are healthier than people who don't" is a logical argument, but it isn't exactly a revelation. Nor can such a statement disprove the science showing that seed oils are safe—so safe that the American Heart Association *recommends* people use plant oils (including seed oils) over other kinds of oils.[141]

It's also interesting to note that two of the primary studies that get cited in support of the argument that canola oil causes inflammation were studies done on rats that were fed an exclusive diet of canola oil or canola oil and salt.[142] Now, I understand that this is done to isolate canola oil as a variable and streamline the researchers' ability to discern cause and effect, but I find it astonishing that anyone thinks these studies reflect anything remotely like a normal human diet. Of course, it stands to reason that anyone who goes on an all-canola oil diet for weeks on end is going to have some ill effects. Any single-food diet for a month straight would turn your stomach upside down.

Before we move on to the next criticism of canola oil, I want to point out that the research shows canola oil is also an excellent source of vitamin E and vitamin K. Vitamin E is an antioxidant, meaning it protects cells in the body from damage caused by exposure to environmental pollutants

---

141 Lichtenstein, A., et al. "2021 Dietary Guidance to Improve Cardiovascular Health: A Scientific Statement From the American Heart Association." *AHA/ASA Journals.* 2 November 2021: 144(23). https://doi.org/10.1161/CIR.0000000000001031

142 Mboma J, et al. "Effects of Cyclic Fatty Acid Monomers from Heated Vegetable Oil on Markers of Inflammation and Oxidative Stress in Male Wistar Rats." *Journal of Agricultural and Food Chemistry.* 2018 Jul 11;66(27):7172-7180. doi: 10.1021/acs.jafc.8b01836. Papazzo A, et al. "The effect of short-term canola oil ingestion on oxidative stress in the vasculature of stroke-prone spontaneously hypertensive rats." *Lipids in Health and Disease.* 2011 Oct 17;10:180. doi: 10.1186/1476-511X-10-180.

(also called "free radicals"), and helps with immune function,[143] and vitamin K, which can also be found in leafy greens and fermented foods, is important for bone health and blood clotting.[144] Other vitamins and minerals in canola contribute to its anti-microbial, anti-inflammatory, anti-obesity, anti-diabetic, anti-cancer, neuroprotective, and cardioprotective qualities.[145]

Regarding the question of safe consumption, sometimes while harvesting I will chew on the seeds because they taste quite good! So, I know firsthand they are safe to ingest in reasonable amounts—though I am not suggesting you should make it a regular habit to eat a whole field full!

Canola also has low levels of glucosinolates, which are an active compound also found in broccoli and radish. Glucosinolates contribute to regulatory functions in inflammation, stress response, metabolism, antioxidant activities, and antimicrobial properties, and they trigger enzyme activity to protect your cells from damage.[146] A recent study by the USDA found that glucosinolates "contribute to the reduced development and growth of several diseases, including cancer."[147] So, we see that compounds in canola can actually help in the fight against disease and cancer.

One of the other growing concerns around all cooking oils, not just canola, is the "smoke point." Different oils have different smoke points, meaning the temperature at which the fats begin to break down and the oil begins to smoke. This is of concern because not only does heating oil past its smoke point release unpleasant odors and make food taste

143 "Vitamin E: Fact Sheet for Health Professionals." *National Institutes of Health, Office of Dietary Supplements.* https://ods.od.nih.gov/factsheets/VitaminE-HealthProfessional/

144 "Vitamin K: Fact Sheet for Health Professionals." *National Institutes of Health, Office of Dietary Supplements.* https://ods.od.nih.gov/factsheets/vitaminK-HealthProfessional/

145 Shen J, et al. "A Comprehensive Review of Health-Benefiting Components in Rapeseed Oil." *Nutrients.* 2023 Feb 16;15(4):999. doi: 10.3390/nu15040999.

146 Smith, Ashleigh. "What Is Rapeseed Vs Canola." True Leaf Market. 21 November 2023. https://www.trueleafmarket.com/blogs/articles/what-is-rapeseed-vs-canola

147 Ortiz, I.N., et al. "Effects of harvest day after first true leaf emergence of broccoli and radish microgreens." *Technology in Horticulture.* 2024: 4. https://doi.org/10.48130/tihort-0023-0031

burnt but it can also have harmful effects on health. When cooking oils are heated past their smoke points, the fatty acid content of the oil increases, and free radicals are released (those environmental pollutants we mentioned above).

But canola oil has one of the highest smoke points among common cooking oils, 468°F, and high oleic canola oil, which is typically used in commercial frying, has an even higher smoke point of 475°F.[148] High oleic canola oil has more oleic acid, a "heart-healthy" monounsaturated fat, and this makes it even more stable when it comes to high heat and shelf life. For comparison, extra virgin olive oil has a smoke point of only 331°F.[149]

According to Harvard Professor of Nutrition Guy Crosby, "The formation of oxidation products of polyunsaturated fatty acids during prolonged commercial deep-fat frying ... is less of a concern for canola oil than for oils with higher levels of more readily oxidized polyunsaturated fat such corn, soybean, sunflower, and safflower oils."[150]

Similar research and findings were published by a team of researchers led by Dr. Grootveld at the World Health Organization. Following tests of harmful quantities of oxidation products, Grootveld stated unequivocally, "Unless exposed to [high temperature] frying practices ... the authors accept that PUFA-rich culinary oils offer little or no threat to human health. Indeed, unperoxidized, intact essential FAs therein such as linoleoyl- and especially α-linolenoylglycerols offer valuable protective health benefits."[151]

As we've seen before, well-meaning people can still fall victim to junk science based on poor methodology, deductive reasoning, broad

---

148 "Canola Oil" *Canola Oil Council of Canada.* 2024. https://www.canolacouncil.org/about-canola/oil/

149 Ibid.

150 "Ask the Expert: Concerns about canola oil." *The Nutrition Source, Harvard T.H. Chan School of Public Health.* 13 April 2015. https://nutritionsource.hsph.harvard.edu/2015/04/13/ask-the-expert-concerns-about-canola-oil/

151 Moumtaz, S., *et al.* "Toxic aldehyde generation in and food uptake from culinary oils during frying practices: peroxidative resistance of a monounsaturate-rich algae oil." *Scientific Reports.* 2019: 9, 4125. https://doi.org/10.1038/s41598-019-39767-1

extrapolations from small, unrepresentative sample sizes, or cherry-picked data. Curiously, many anti-seed oil advocates, including a well-known authority on nutrition and wellness who has published multiple best-selling books, often quote partial excerpts of Grootveld's research but then omit his conclusions that don't fit with their beliefs. They cherry-pick elements of the study that fit their misguided theories while ignoring the overall conclusion that seed oils are safe *and* offer "valuable protective health benefits." Whether this is an innocent misunderstanding in a misreading of the science or a deliberate attempt to fear-monger (and persuade the public into buying an expensive subscription diet plan) is not for me to say. But I do believe if someone is going to advise people on matters of health and diet, they should at the very least be honest and accurately present the evidence they use to back up their claims.

To return to our Tylenol metaphor, take two and you'll cure a headache, take ten and you will damage your liver! We must follow the instructions for safe use. If people consistently heat canola oil—or any oil—past its smoke point, or they reuse cooking oil more times than is recommended, then they aren't following the guidelines for safe use. But that is user error—not an inherent risk of the product. Don't heat your oil so high it starts to smoke or burn.

So, if all oils have smoke points, and canola oil has one of the highest smoke points among cooking oils, why do people single out canola oil as "toxic?" Again, I can only surmise that the people pushing this narrative benefit from creating anxiety around the safety of our food supply because they have an alternative or "cure" to sell you.

Doubling back briefly, I want to touch on the other advantage of canola oil's stability: high oleic canola oil. Not only does this make for better, safer frying but it is also an advantage when it comes to packaged foods like potato chips. Because the bonds between fats in canola oil are so strong, they last longer, and that means a better shelf life for any products that use canola oil as their source of fat.

This also gives us some insight into health claims against canola oil. Because it is such a good oil—from the perspective of health, shelf-life, and cost—it is used in many packaged goods or fried foods, which can lead people to associate canola oil with "unhealthy" or "junk" food. I don't like to assign moral value to any food, but it's also clear that canola oil is not at fault here for negative health outcomes like weight gain or obesity. The culprit is clearly the overconsumption of foods high in carbohydrates, sugar, and sodium.[152]

Shelf life also has an impact on cost. High oleic canola oil can be heated multiple times, so companies that use the product commercially—think any fast-food joint that sells millions of pounds of French fries daily—are getting better value for money. It has a longer shelf life, which means companies that use the oil in packaged goods have more opportunities to sell their product. And for consumers, it means the food you buy is less likely to end up as food waste by going rancid.

Now, for whatever reason, a long shelf life is a red flag to some consumers. It seems "unnatural" for food to last on the shelf. It should be going rotten like an apple fallen from the tree! The stability of canola oil is due in part to the process by which it is refined—another reason critics claim it is toxic.

So, how is canola oil made? Canola oil is part of a family of cooking oils, sometimes referred to as RBD oils or refined, bleached, and deodorized oils, which undergo a process of refinement that has been in use in the industrial production of oils since the 1930s. This process, by the way, is the same for all cooking oils unless they are labeled "unrefined," "cold-pressed," or "extra-virgin," which of course you'll pay a premium for.

One concern about the refining process often voiced by naysayers is the use of a solvent called hexane. When canola seeds are pressed, they form a cake and become saturated with hexane, which assists with

---

152 "Scientists debunk claims of seed oil health risks." *Harvard T.H. Chan School of Public Health.* 22 June 2022. https://www.hsph.harvard.edu/news/hsph-in-the-news/scientists-debunk-seed-oil-health-risks/

extracting the remaining oil, after which the hexane is separated and removed from the oil.

There are critics of the process who try to make it sound horribly complicated and full of high heat and chemicals, but that is simply false. Another fear may be because when canola is filtered it is called "bleaching." That term makes one believe bleach is added! This is also incorrect. The process uses a special clay that filters the oil and removes any particles from the oil. This gives canola oil its aesthetically appealing golden color.

The concern stems from whether there is residual hexane in the cooking oils that go to market and are later ingested. However, studies examining residual hexane in food products have shown that the levels across all samples were well below our old friend the maximum residue level (MRL).[153] According to experts, "There is no evidence to substantiate any risk or danger to consumer health when foods containing trace residual concentrations of hexane are ingested."[154]

Indeed, you are exposed to higher levels of hexane when you fill up your car with gasoline.[155] Does that mean you're going to stop topping up your tank or driving your car?

Claims that canola is hazardous because it is a GMO or highly refined, or that omega-6 fatty acids cause inflammation, are all bogus. They do not stand up to the science. But I want to take it a step further and show you that not only is canola oil *not harmful*, but it is also actually beneficial in many ways.

We have touched on some of the health advantages of canola oil, like its composition of heart-healthy fats and essential vitamins, its positive impact on cholesterol levels, and its ability to reduce the risk of heart

---

153 Cravotto C, et al. "Towards substitution of hexane as extraction solvent of food products and ingredients with no regrets." *Foods.* 2022;11(21):3412. doi: 10.3390/foods11213412.

154 Swanson, R. G., Regents Professor. "Hexane Extraction in Soyfoods Processing." *Department of Food Science, Washington State University.* 2009.

155 Crosby, Guy. "Ask the Expert: Concerns about canola oil." *The Nutrition Source, Harvard T.H. Chan School of Public Health.* 13 April 2015. https://nutritionsource.hsph.harvard.edu/2015/04/13/ask-the-expert-concerns-about-canola-oil/

disease. These qualities are why canola oil has been named a "heart smart" oil by multiple health organizations, including the US Food and Drug Administration.[156]

For decades, countless peer-reviewed clinical trials involving thousands of human volunteers have repeatedly shown that canola oil can help lower blood cholesterol levels and reduce the risk of coronary heart disease, cancer, diabetes, and high blood pressure.[157] A meta-analysis of over 40 clinical trials investigating the effects of canola oil found that it "significantly improved different cardiometabolic risk factors compared to other edible oils," including improvements in glycemic indices, inflammation, and blood pressure.[158] And a growing body of evidence indicates canola oil has positive effects on several chronic health problems, including heart disease, diabetes, and metabolic syndrome.[159]

There are NO scientific peer-reviewed studies that link an average consumption of canola oil to chronic disease. There is only conjecture and fearmongering. I have seen some authors claim that higher cardiovascular disease (CVD) rates stem from toxins in overheated vegetable oil, but according to the American Heart Association, death rates from CVD have declined by 60 percent since 1950.[160] I have also seen unsubstantiated claims that there is a greater possibility of cancer, particularly colon

---

156 Johnson, Gary. "Qualified Health Claims: Letter of Enforcement Discretion - Unsaturated Fatty Acids from Canola Oil and Reduced Risk of Coronary Heart Disease." *U.S. Department of Health and Human Services.* 14 November 2017.

157 "Canola Oil" *Canola Oil Council of Canada.* 2024. https://www.canolacouncil.org/about-canola/oil/

158 Amiri M, et al. "The effects of Canola oil on cardiovascular risk factors: A systematic review and meta-analysis with dose-response analysis of controlled clinical trials." *Nutrition, Metabolism, and Cardiovascular Diseases.* 2020 Nov 27;30(12):2133-2145. doi: 10.1016/j.numecd.2020.06.007.

159 Lin, L, et al. "Evidence of health benefits of canola oil." *Nutrition Reviews.* June 2013;71:370-385. https://onlinelibrary.wiley.com/doi/epdf/10.1111/nure.12033

160 Martin, S., et al. "2024 Heart Disease and Stroke Statistics: A Report of US and Global Data From the American Heart Association." *AHA/ASA Journals.* 2024: 149(8). https://doi.org/10.1161/CIR.0000000000001209

cancer (as one would expect if the colon were exposed to "toxins" from burnt oil). Yet colon cancer diagnoses have declined since 1992.[161]

I also want to point out that many anti-seed oil claims are based on correlation. Correlation is not causation. Correlation is seeing the death rate from heart attack and stroke in men in Norway drop precipitously after 1939 when Germans occupied the territory and took all the livestock.[162] Norwegians were forced onto a vegetarian diet, and the deaths from heart attack and stroke plummeted between 1940 and 1945. Is it because they stopped eating meat, or is it perhaps because there was a more significant, relevant, and immediate cause of death in men during WWII?

This is an example of correlation potentially mistaken for causation. Just because A occurs and then B occurs does not mean that A caused B! In this case, we've got the question of why heart attacks decreased during WWII in Norway. Did the lack of meat in the diet cause that reduction in heart disease or did it simply happen at the same time? Without further proof, there is no way of knowing. Which is why I reiterate, don't confuse causation and correlation.

Health and disease are multifactorial issues—genetics, lifestyle, diet, access to preventative care, testing, diagnosis, health care, treatment availability, public awareness, research funding, and even environment are all contributing, overlapping factors. Should theories be studied and researched scientifically? Of course, and we should continue to do everything we can to understand food and nutrition, especially as they relate to health outcomes. That is why I look for peer-reviewed conclusions and not flimsy theories or dubious correlations when I make my choices. I hope you will, too!

---

161 "Key Statistics for Colorectal Cancer." *American Cancer Society*. 29 January 2024. https://www.cancer.org/cancer/types/colon-rectal-cancer/about/key-statistics.

162 Axel Strøm, R.et al. "MORTALITY FROM CIRCULATORY DISEASES IN NORWAY 1940-1945."
The Lancet. 1951: 257(6647), 126-9. https://doi.org/10.1016/S0140-6736(51)91210-X.

To return to the benefits of canola, there are also environmental advantages to growing canola. First, growing, producing, and processing canola and canola oil are largely domestic. We are growing the crops here in North America, and our capacity to process canola has significantly increased in the last ten years as we've built more of the "crush facilities" where processing takes place. That means more jobs in North America as opposed to having to ship raw seeds to China, Japan, or Europe for processing and then importing the cooking oil back at a higher cost.

Additionally, we aren't cutting down trees and clearing forests to open new land for crops. We are growing canola on existing farmland—often family-owned, legacy farms like mine. So, unlike production for palm or coconut oil, canola production involves no deforestation[163]—and perhaps more critically, there is no concern about unethical practices like child labor or slave labor,[164] which exist in less regulated markets and developing countries. And I won't even get into the monkey slave labor used in the production of some coconut oil.[165]

Lastly, the fryer isn't the end of the line for canola oil. Much of the canola oil used commercially in North America gets a second life as biodiesel, a sustainable, eco-friendly fuel source.[166] Canola biodiesel is also being used at petroleum refineries to lower the greenhouse gas emissions of transportation fuels.[167] It is a cleaner-burning alternative to petroleum fuels and has an important role to play in lowering carbon footprints.

---

163 Meijaard, E, et al. "The environmental impacts of palm oil in context." *Nature Plants.* 6:1418–1426 (2020). https://doi.org/10.1038/s41477-020-00813-w

164 "Coconut Controversy: The Heartbreaking Truth Behind Your Coconut Products." *Food Empowerment Project.* https://foodispower.org/our-food-choices/coconut-controversy/

165 Fobar, Rachel. "Monkeys still forced to pick coconuts in Thailand despite controversy." *National Geographic.* 19 February 2021. https://www.nationalgeographic.com/animals/article/monkey-labor-continues-in-thailands-coconut-market

166 Ge, Jun Cong & Yoon, Sam & Choi, Nag. "Using Canola Oil Biodiesel as an Alternative Fuel in Diesel Engines: A Review." *Applied Sciences.* 2017;7:881. 10.3390/app7090881.

167 "Biodiesel." *U.S. Canola Association.* 2024. https://www.uscanola.com/biodiesel/biodiesel/

So, what is the truth about canola? The truth is that canola is a safe, healthy option for cooking oil, provided it is handled according to safe use practices. There are heart-healthy benefits to cooking with canola oil, and there are additional advantages to canola when it comes to the environment, the economy, and labor practices. Remember that the next time you drive past those beautiful, bright yellow canola fields.

Chapter 12

THE PROTEIN PROBLEM

Vegan and vegetarian diets are becoming increasingly mainstream and popular. People adopt these diets for various reasons, and many believe that they are saving the planet by eliminating meat from their diet. I am not here to judge anyone for their personal choices when it comes to what they eat, but the truth is that our entire agricultural food system would be at serious risk if we lost all the protein that livestock provides. Meat is a key source of protein to feed the world, and I think it is important for people to understand that the choice to be vegan or vegetarian categorically cannot work on a global scale.

The main reason a meat-free diet is unfeasible on a global scale is that humans need protein. Protein builds and repairs tissues, regulates metabolic processes, maintains proper pH, supports the immune system, and serves many more critical functions like transporting and storing nutrients and providing us with energy. Proteins are the building blocks of life. Indeed, the increased consumption of meat in our diets 1.5 million years ago led to evolutionary changes— for one, eating meat made it possible for us to evolve larger, more complex brains.[168] Meat made us human. By the way, nature didn't give us incisor teeth for tearing lettuce apart! Lol.

The amount of protein we should eat per day depends on factors like age, sex, height, and weight, but on average it works out to about 45 grams for women and 55 grams for men. That equates to about two portions of meat per day.

---

168 Ireland, Corydon. "Eating meat led to smaller stomachs, bigger brains." *The Harvard Gazette.* 3 April 2008. https://news.harvard.edu/gazette/story/2008/04/eating-meat-led-to-smaller-stomachs-bigger-brains/

There *are* plant-based sources of protein, so if we were to eliminate meat as a source of protein, we could hypothetically consume more plant protein to offset the loss. So, why don't we do that? Because it would be physically impossible and financially unviable.

It is estimated that about four or five percent of Americans identify as vegetarians and about two percent as vegans.[169] That figure is roughly the same as the global average: according to a survey conducted across 28 countries, about five percent of the world's population is vegetarian.[170] The percentage varies depending on cultural context—for example, India has an unusually high percentage of vegetarians (between 22 and 33 percent), in large part because of Indian religious practices.[171] But to meet the dietary needs of the other 95 percent of the global population—roughly 7.8 billion people—we would need to *drastically* increase plant-based food production, to say nothing of the labor required in such an undertaking. We simply do not have the land capacity to do that.

Let's break it down into numbers. Most of the Earth's surface is water—a whopping 71 percent. That leaves 29 percent as land. So, as we've discussed before, land is a finite resource. About 38 percent of the total land on Earth is currently used for agricultural purposes.[172] Agricultural land can be divided into two categories: arable land, where we grow plants for food, and marginal land, which is not suitable for growing food crops. Arable land makes up just 33 percent of agricultural land,

---

169 "Nutrition and Food." *Gallup.* 2024. https://news.gallup.com/poll/6424/Nutrition-Food.aspx

170 "An exploration into diets around the world." *Ipsos Mori.* August 2018. https://www.ipsos.com/sites/default/files/ct/news/documents/2018-09/an_exploration_into_diets_around_the_world.pdf

171 Ibid.

172 Anderson, Sharissa. "Cattle and Land Use: The Differences between Arable Land and Marginal Land and How Cattle Use Each." *Clarity and Leadership for Environmental Awareness and Research at University of California, Davis.* 17 January 2023. https://clear.ucdavis.edu/explainers/cattle-and-land-use-differences-between-arable-land-and-marginal-land-and-how-cattle-use

which means that less than 13 percent of all land on Earth is suitable for growing plants for food.[173]

The switch to a meat-free diet for everyone would mean we would need to feed 100 percent of people using just 13 percent of the land. The numbers don't add up.

Marginal land, on the other hand, can be untouched and undeveloped, or it might be pastureland—meadows, prairie, mountains, or other terrain where growing crops would be impossible or so difficult it wouldn't be worth the effort. But a cow or a sheep is happy to amble through pastureland grazing, napping, and doing whatever else cows and sheep do.

As we've seen in previous chapters, however, not all soil is created equal. Different geographical regions and even different plots of land within the same area can have vastly variable soil compositions. Some might have more clay or sand, others will be rocky, and still other areas will be rich and fertile. Some soil is not "arable" or productive for agriculture.

So, what do you do with land that cannot be used to grow crops? Land is a limited resource, so we need to use it efficiently. If land can't be farmed for crop production, the next best thing in terms of food production is to use it for grazing. Land that isn't suited for crops can make excellent ground for native grasses, which can then be used to graze cattle (or sheep or goats, depending on where you live in the world). Cows are incredibly efficient at converting cellulose (the fiber found in plants) into high-quality protein for human consumption. Cows, in essence, are sun-changers! They take the energy from the sun that grows the cellulose and convert the sunlight into high-quality protein! What a wonderful gift.

More importantly, cows can help restore healthy soils, conserve sensitive species, and enhance overall ecological function. Cattle ranchers have a vested interest in maintaining soil health, just as farmers do—it is our most important resource, after all. If ranchers overgraze and let the soil degrade, it can affect the health of their cattle and have a negative impact

---

173 Ibid.

on their long-term ability to graze those lands. Farmers and ranchers are the best environmentalists in the world because we have to be. It is not just our privilege to protect the land; it is an obligation.

Some livestock are grazed on arable land, which prompts people to assert that we could take back that land and use the plants that animals eat to feed people instead. There are several problems with this proposition. Livestock production globally uses crops grown on about 40 percent of arable land; however, this does not mean that animals are eating 40 percent of our food.[174]

First, we know from our discussion in Chapter 11 that livestock can eat "feed grains" that do not meet the requirements for human consumption. There are several reasons a grain crop might fail, with drought, disease, and poor quality being the most common. It happens every year to some percentage of crops. But beef cows, hogs, and poultry can safely eat these foods, essentially recycling grains that would otherwise go to waste.

Critics will also argue that by feeding cattle, we are wasting food that could be used to feed people, but that is frankly incorrect. Obviously, we can't eat grass—we all know that much. But what about the grain that is fed to cows as a supplement? There are two main reasons why we feed grain to cows. The first is that in winter in many parts of the world, including Canada where I live, you aren't going to find a blade of grass anywhere for cows to eat. So, we feed them grains like barley, which is high in protein and nutritionally dense thanks to its starchy kernels. This high-energy food keeps them warm and fuels them better than a diet of just hay. A cow would need to eat far more hay or grass to get the same protein and nutrients they get from grains.

The other reason we supplement with grains is that because of the evolutionary adaptations of a cow's stomach, they can eat things we can't. Grass, for one, but also wheat and other grains that have been frozen, which we as humans could, therefore, no longer use to make flour. Wheat

---

174 Ibid.

shrivels up and turns black at the first sign of frost, and the frost kills the crop before it is ripe. But cows don't mind at all. It's the same with the barley we mentioned above. Cows are fed the barley that isn't suitable for making beer or barley syrup.

Lastly, cows are only fed grain in their last 90 days or so. It accounts for a small portion, about 14 percent, of their overall diet. Again, because grain is a high-energy, high-protein food source, it helps build fat in the cow. This is good for keeping the cow warm through winter, and it's good for consumers down the line who want some fat marbling in their meat for flavor. As an aside, this is why labels like "100% grass-fed beef" are just another ploy to get a little deeper into your pocketbook. All cows are grass-fed for most of their lives.

So where does most of the diet for livestock animals come from? A whopping 86 percent[175] comes from foraging or agricultural byproducts like beet pulp, dehulled oats, canola meal, molasses, and feed grain. It also includes "crop residues" like straws, stems, and tops of plants left over after harvest, as well as "spent grains" from breweries and distilleries. In the US, 93 percent of beef cattle's diet consists of food that is inedible to humans.[176] When canola is processed into oil, for example, about 42 percent of the resulting product is oil while the other 58 percent is canola meal—which is great for animal feed and not much else. Including animals in agricultural production enables us to sustain a food system with little to no waste.

Cows are like nature's recycling bins. Their rumen (the largest stomach compartment in a cow) is basically one big fermentation tank. They even eat other food waste from farms, like sugar beet pulp, corn stalks, and potato peels that would otherwise end up in a landfill.

The resulting loss affects the financial and ecological resources invested to grow the grains that would have also gone to waste.

---

175 Ibid.
176 Ibid.

Thanks to the meat industry, a farmer can still squeeze out some profit from a crop that is unfit for human consumption but perfectly safe for animals (who have different digestive systems, suited to such sources of energy). This means the farmer does not take a total loss. The farmer can stay in business, and the cows, chickens, or pigs will convert that shriveled-up wheat into high-quality protein like meat, eggs, and dairy. It's a safe, efficient way to use up feed grains that would otherwise go to waste.

So, now, we have a better understanding of livestock's relationship to arable land. But what about the other two-thirds of agricultural space, the marginal land? Beef cattle in particular come under fire for the amount of land used to raise and graze them—but very few people put that land use in context.

Most cattle are raised on marginal land that isn't suited to other purposes, like growing crops or urban/suburban development. This is land that receives little to no water, has low-quality soil, has high mountains with steep slopes, and is barren or rocky. Do you want to try your hand at growing wheat on the side of a mountain in the Canadian Rockies? Good luck to you—you'll need it. But cows have adapted to navigate rough terrain and survive in harsh climates.

Furthermore, as we discussed earlier, it benefits cattle ranchers to practice sustainability. Ranchers who understand how to raise cattle properly avoid overgrazing because we know the land is a critical resource. If cattle overgraze an area, the grass will start to die. Many ranchers therefore practice "rotational grazing," moving cattle from one plot to another every few weeks. This also helps protect the soil and prevent erosion. Ranchers need to protect the ecosystem if they want to keep grazing cattle on a particular plot. It's as simple as that.

Meanwhile, as the cattle are grazing, they are also recycling the nutrients from the grass. They eat the grass and ferment it and convert some of it to protein—and then, of course, cattle manure comes out the other end and goes back to the land to fertilize the grass. Cows also act as natural weed killers because they graze on weeds, which gives other plants, including native species, a chance to compete for nutrients and resources. This is how cattle are used to restore depleted soil and land! What could be more natural than that?

This brings us to the last point I want to make about cattle farming specifically, which is the role cows play in carbon sequestration. Does that surprise you? Most of the time when people think about cows and emissions, they think about the methane gas that cows release as part of their bodily functions. So, how does it work? How do cows help sequester carbon?

We know that plants pull carbon dioxide (CO_2) from the air and store it in their roots, stems, and leaves, or in the soil. Also, it's true that cows eat plants, and whatever carbon from the plants they don't use is released through belching and in their manure as methane (CH_4). But that isn't the end of the cycle.

After about ten years, the methane gas gets converted back into CO_2, which plants can then reabsorb. Even better, the methane in manure provides an energy source for microbes and organic material in the soil. Soil microbes recycle organic matter either by converting carbon for long-term storage or by releasing carbon as CO_2 in the soil for plants to use in photosynthesis—both of which are forms of carbon sequestration. This is called the "biogenic carbon cycle," and it shows us that cattle do not emit new carbon into the atmosphere; rather, they are part of a natural form of carbon recycling.[177]

Land use and environmental claims aren't the only reasons people consider not eating meat. There is no shortage of claims about the health benefits of a vegan or vegetarian diet, but almost none of those claims stand up to scrutiny. Popular media, films, and books don't just claim that eliminating meat from our diet is "healthier," but also that it can

---

177 Ibid.

prevent, cure, and reverse all manner of chronic diseases, from dandruff to heart disease and cancer.[178]

If it sounds too good to be true … well, it probably is. Ask any dietician or nutritionist and they will tell you that a vegetarian option isn't automatically a healthy choice. Mock meats like fake "chicken" nuggets, veggie burgers, and soy hot dogs—a multi-billion-dollar industry— do contain protein, but they are often calorie-dense, loaded with salt, and high in saturated fat.

For example, if you compare Burger King's Whopper with its vegan option, the Impossible Whopper, you might be surprised to find there are only negligible nutritional differences between the two.[179] Both provide about 30 grams of protein and 60 grams of carbohydrates. They have roughly similar calorie counts (657 versus 629) and fat (40 versus 34), but the beef Whopper actually has about 10 percent less sodium. If you offer me a juicy all-beef cheeseburger or a "textured vegetable protein" patty with vegan cheese, I know which one I'm choosing.

Additionally, many plant-based proteins are not "complete proteins," meaning they lack one or more of the nine essential amino acids. Again, we arrive at the protein problem, which is that not all proteins are created equally. A food is considered a complete protein when it contains all nine of the essential amino acids that our bodies can't produce. There are 20 amino acids in total; our bodies can naturally produce eleven of them, but not the other nine.[180] Most vegetarian and vegan proteins, like legumes, nuts, seeds, whole grains, and vegetables, are incomplete proteins. There are some plant-based complete proteins, like chickpeas, quinoa, and tofu,

178 There are multiple outlets and media that have made such claims, including the documentary *Forks Over Knives*, https://www.forksoverknives.com.

179 Lenhoff, Alex. "Beef Whopper Vs Impossible Whopper: Which is healthier?" *Fast Food Nutrition*. 20 August 2019. https://fastfoodnutrition.org/news/beef-whopper-impossible-whopper-which-healthier-1566337500

180 "What are complete proteins?" *Cleveland Clinic*. 6 December 2022. https://health.clevelandclinic.org/do-i-need-to-worry-about-eating-complete-proteins

but you would need to carefully plan your diet to make sure you weren't missing out on essential nutrients.

So, let's set aside the fact that there isn't enough land to produce enough plant protein to properly feed and nourish everyone, and let's imagine everyone decides to become vegetarian. While it is *possible* to substitute plant protein for meat, there are still practical obstacles to adopting a vegetarian diet, such as the level of resources, culinary skills, dietary culture and habits, and nutritional awareness and knowledge.[181] Sometimes, it's as simple as saying not everyone likes tofu!

Furthermore, food sensitivities, intolerances, and allergies—for example, to foods containing gluten, soy, or pea protein—mean that a meat-free diet is not suitable for everyone. This is especially true for people with specific dietary needs, like infants and young children, pregnant women, seniors, and people with chronic diseases, as well as people in low- and middle-income countries who may have limited access to food sources generally.[182]

For example, people who decrease their intake of red meat tend to develop iron deficiency as a result.[183] In fact, 20 percent of women of reproductive age in the US are iron deficient,[184] which can lead to anemia, decline in cognitive function, and negative health effects on any children they may have.[185]

181 Leroy F, et al. "The role of meat in the human diet: evolutionary aspects and nutritional value." *Frontiers in Animal Science.* 2023 Apr 15;13(2):11-18. doi: 10.1093/af/vfac093.

182 Ibid.

183 Sun, H. and C.M. Weaver. "Decreased iron intake parallels rising iron deficiency anemia and related mortality rates in the US population." *Journal of Nutrition.* 2021;151:1947–1955. doi: 10.1093/jn/nxab064.

184 Stevens, G.A., et al. "Micronutrient deficiencies among preschool-aged children and women of reproductive age worldwide: a pooled analysis of individual-level data from population-representative surveys." *The Lancet Global Health.* 2022;10:e1590–e1599. doi: 10.1016/S2214-109X(22)00367-9.

185 Murray-Kolb, L.E.J., and J.L. Beard. "Iron treatment normalizes cognitive functioning in young women." *American Journal of Clinical Nutrition.* 2007; 85:778–787. doi: 10.1093/ajcn/85.3.778.

These considerations help explain why most vegetarians—about 84 percent—eventually go back to eating meat.[186] It isn't sustainable, practical, or healthy, even for people who *want* to give up meat.

In contrast, meat gives us high-quality protein with various nutrients that are readily "bioavailable," meaning we have evolved to easily digest and absorb them. Beef, for example, provides about 20 to 24 grams of protein per 100 grams of meat, and it is a complete protein, meaning it provides all the essential amino acids, fatty acids, and macro- and micronutrients.[187]

In fact, even though meat constitutes only about seven percent of all food consumed globally, it provides approximately 21 percent of the world's protein.[188] It also over-delivers on energy and several micronutrients, especially B vitamins and vitamin A (which we discussed in Chapter 4). That is a huge contribution in terms of ensuring an adequate supply of dietary protein, which is central to addressing global food security and nutrition.

A systematic review of epidemiological clinical studies found that "When meat consumption is part of healthy dietary patterns, harmful associations tend to disappear, suggesting that risk is more likely to be contingent on the dietary context rather than the meat itself."[189] So if you eat nothing but Big Macs all day, every day, you probably won't feel great. But if you follow the mantra of "everything in moderation," then

---

186 Herzog, Hal. "84% of Vegetarians and Vegans Return to Meat. Why?" *Psychology Today.* 2 December 2014. https://www.psychologytoday.com/gb/blog/animals-and-us/201412/84-of-vegetarians-and-vegans-return-to-meat-why

187 Anderson, Sharissa. "Cattle and Land Use: The Differences between Arable Land and Marginal Land and How Cattle Use Each." *Clarity and Leadership for Environmental Awareness and Research at University of California, Davis.* 17 January 2023. https://clear.ucdavis.edu/explainers/cattle-and-land-use-differences-between-arable-land-and-marginal-land-and-how-cattle-use

188 Smith, N.W., et al. "Modeling the contribution of meat to global nutrient availability. *Frontiers in Nutrition.* 2022; 9:766796. doi: 10.3389/fnut.2022.766796.

189 Johnston, B., et al. "Non-communicable disease risk associated with red and processed meat consumption—magnitude, certainty, and contextuality of risk?" *Animal Frontiers.* April 2023; 13(2):19–27. https://doi.org/10.1093/af/vfac095

meat is absolutely part of a healthy, balanced, nutritious diet. And any dietician, doctor, or nutritionist will tell you the same thing.

Another major factor in the meat debate is cost and access. Notably, arguments for a widespread reduction in meat consumption come mostly from high-income countries to the detriment of low-income countries.[190] One study found that such efforts to eliminate or reduce meat consumption could impede the progress being made toward reducing malnutrition and its cognitive and physical impacts and stifle economic development, especially in the global South.[191] Meat is a vital source of protein, calories, and nutrients for billions of people who struggle to get enough to eat.

My last point may sound silly given the gravity of what's at stake, but I want to highlight that even if people stopped eating meat, we would still need some meat production for our pets. Cats and dogs *need* animal protein, just like us. And we're lucky that our pets don't have the same hang-ups about what part of the cow or chicken they're eating. To them, meat is meat, and protein is protein. They evolved (just like us) to be able to digest and make use of all parts of an animal. That means meat byproducts that we would otherwise discard don't go to waste. They get ground up and fed to our cats and dogs. It doesn't sound nice, but your pet loves it.

Whether you eat meat or only certain kinds of meat or no meat at all is irrelevant. Eat what you like; I don't begrudge anyone that choice. But the fact remains that we need domesticated livestock to help feed the world. If we remove meat from the global food supply, we are removing essential sources of protein and nutrients for billions of people.

We would also lose the critical ecological function these animals serve in their environments, and we would be left with countless tons of food waste from agricultural byproducts. Well-managed livestock converts substantial quantities of nonedible plant material into protein, recycles

---

190 Leroy F, et al. "The role of meat in the human diet: evolutionary aspects and nutritional value." *Frontiers in Animal Science.* 2023 Apr 15;13(2):11-18. doi: 10.1093/af/vfac093.
191 Ibid.

nutrients back to the land, improves soil health, prevents erosion, sequesters carbon, and benefits its ecosystems in countless other ways.

Overall, the efficiency of protein production by mankind has advanced by leaps and bounds, enabling us to take on the challenge of feeding the global population. I'm proud that we as farmers have helped make that progress. Can we get better? Yes. We are always trying to find ways to improve our systems and reduce our impact.

Ultimately, like so much of what we've discussed, the conversation boils down to this: be careful what you wish for. Livestock animals have played key roles in their ecosystems for millions of years. Right now, our system is balanced, and trying to end meat production would completely and utterly upset that balance, with grave consequences for some of the world's most vulnerable people.

Chapter 13

WHAT FOLKS DON'T UNDERSTAND ABOUT FARMING

Farming isn't flashy—unless your idea of flashy includes flannel, overalls, and heavy machinery. You won't find us in the headlines or trending on TikTok. Most farms in North America are family-owned, as we've said. We are "stay-ups," not start-ups. We've been around too long to be the next new thing. We plod along behind the scenes, doing the work of growing food to feed the world, and because of that our stories, our work, and the care that we put into what we do, we can go unnoticed and be misrepresented.

There is a growing divide between farmers and consumers, and I believe we can bridge that gap by communicating better and sharing our knowledge and experience. It's why I've written this book. To be fair, many farmers only talk to other farmers, so I'm not looking to place blame on anyone for the way things stand. However, I do think it is important that people understand the value farmers and food producers add to society and the painstaking efforts we take to protect the environment.

I recently drove through an area with lots of old farms. When people like my great-grandfather moved here to break ground, the first thing they would do was build a log house, and the second thing was plant trees around the house as wind barriers. Most farmyards—even today—have trees all the way around, many of them likely planted by the original farmers. It got me thinking about what it feels like to plant a tree.

When you plant a tree, you are doing something for the generations ahead. You plant it knowing that you may not be around in 30 years' time to see its beauty, enjoy the shade of its branches, or watch it grow to its full potential. If you've ever planted a tree, maybe you know that feeling.

I say this because I think the feeling of planting a tree is the same feeling we have as farmers. We plant the seeds and harvest the crops, but we won't ever see the bread on the table. The food we grow is primarily for other people. I'm really proud that I get to do that, but there is more to it than a warm, fuzzy feeling. Farmers, especially industrial farmers, provide good, reliable jobs for people.

According to the USDA, over 22 million jobs were related to agriculture and food production in 2022.[192] That's equivalent to 10.4 percent of total employment for the entire country. And in 2023, American farms generated $204.5 billion toward the US gross domestic product (GDP).[193] The sector as a whole was worth over $1.5 trillion.

However, those figures can seem somehow meaningless to the people I meet at networking events for entrepreneurs and innovators. These are people who have built and sold a dozen different multimillion-dollar startups, but who only employ one or two people—likely virtual assistants or programmers from overseas who are working for low incomes. They don't have any idea what it would take to stay in business for 40 years. And maybe they even see it as a failure that I employ so many people—that's more money out the door.

It takes passion, perseverance, and vision to be a farmer. And it takes a certain amount of loyalty to your people—not just those you employ, but also the people who depend on you to keep doing your job so that they, in turn, can keep feeding themselves and their families. Farms aren't startups; they are usually generational, legacy businesses. And they are hard work.

---

192 "Ag and Food Sectors and the Economy." *U.S. Department of Agriculture, Economic Research Service.* 19 April 2024. https://www.ers.usda.gov/data-products/ag-and-food-statistics-charting-the-essentials/ag-and-food-sectors-and-the-economy/?topicId=b7a1aba0-7059-4feb-a84c-b2fd1f0db6a3

193 Ibid.

There are costs and risks in farming that people may not realize. We often live in rather desolate locations. There are few places more inhospitable to man than Saskatchewan in winter. In my great-grandfather's time, a serious hailstorm could wipe out entire crops in minutes.

I often think of my great-grandfather breaking 22 acres of rocky soil with a single ox-pulled plow. Imagine an ox pulling a spade through the soil. It is slow going, and the ox does not always cooperate or walk in a straight line. If you hit a rock or a tree stump, then you have to dig it up, sometimes pulling it out root by root.

It's different now, of course, but it still isn't easy. To this very day, we are fighting disease, weather, and pests, so we need every tool in the toolbox to combat them. These are old problems, but we have developed better and better (and safer and safer) means of mitigating them. Refusing to use modern solutions—like GMO plants or fertilizers—makes as much sense as refusing to use modern medical techniques to treat illness. Would you rather a doctor applied leeches to your wound because they come from nature, or would you prefer to be properly stitched up in a hospital and given antibiotics?

There is a similar misunderstanding when it comes to agricultural meat production. We've kept our discussion largely to plant-based farming thus far because that's what I grow, but I also have a background in animal nutrition and health, and no discussion of food production would truly be complete without at least touching on the meat industry.

I want to preface this part of the conversation by saying it doesn't matter what you choose to eat or not eat. That's a personal choice, and it's not my place to pass judgment. I would, however, like to introduce some of the thinking that determines why we graze animals in certain places, and I'd like to explain how it benefits consumers and the environment.

Cattle farming, in particular, tends to come under fire from environmentalists and animal rights activists. In 1971, Frances Moore Lappé wrote *Diet for a Small Planet,* in which she claimed that sixteen ounces of grain can produce sixteen ounces of grain-based food, but only one ounce of beef. Essentially, she was arguing that we should all adopt a plant-based diet for environmental and ecological reasons. Various versions of her argument have become popularized such that today about five percent of Americans consider themselves vegan or vegetarian.[194]

Now, objectors will assert that cows produce methane gas, which contributes to climate change. While it is true that cattle are the top agricultural source of greenhouse gases, that is not reason enough to stop eating beef specifically or meat more generally.[195] First, cows are responsible for only two percent of emissions, and we are steadily reducing their carbon footprint.

Second, much as advances in agricultural technology have empowered farmers to produce higher yields from the same, limited acreage, the same thing is happening with cattle farming. Thanks to advances in genetics, nutrition, and breeding, we can feed more people with fewer cows because each cow produces more meat than was once possible. In the 1970s, 140 million cows were needed to meet demand; today, that number has dropped to 90 million—even though the US population has increased more than 50 percent in that time.[196] That equates to a significant drop in emissions.

Scientists are also developing ways to make cattle farming more sustainable. One solution appears to be introducing seaweed into cows' diets. One study showed a 60 percent reduction in emissions from cows by simply incorporating seaweed into one percent of their diet.[197]

So, to boil this all down—good, arable land produces crops that people can eat, and rough, inarable land produces grasses that cows can eat and convert to quality protein. Just watch a cow walk halfway up a rugged

---

194 Jones, Jeffrey. "In U.S., 4% Identify as Vegetarian, 1% as Vegan." *Gallup.* 24 August 2023. https://news.gallup.com/poll/510038/identify-vegetarian-vegan.aspx

195 Quinton, Amy. "Cows and Climate Change: Making cattle more sustainable." *University of California Davis.* 27 June 2019. https://www.ucdavis.edu/food/news/making-cattle-more-sustainable

196 Ibid.

197 Roque, BM, et al. "Inclusion of *Asparagopsis armata* in lactating dairy cows' diet reduces enteric methane emission by over 50 percent." *Journal of Cleaner Production.* 10 October 2019. Vol. 234, p.132-138.

mountain and convert that unusable grass into high-quality protein, which people in turn can eat! That is nature, it's the cycle of life.

Don't eat beef or any meat if you don't want to. It's a free country, and I respect everyone's right to choose to eat what they like. But understand that eliminating cows from the food supply would mean losing our supply of high-quality protein—including in low-income or developing regions where people may not have any alternative to meat. It also breaks the balance of the cycle wherein inarable land can still be used to produce food and waste material gets turned into nutritious food for consumers.

One in ten people in the world doesn't get enough calories each day and raising cattle for beef is one way of addressing that deficit. Whatever your personal choice may be, I don't think it's good or right or fair to take the option away from other people who may really need that food source.

As farmers, we do everything we can to work *with* nature when we can, because we spend so much of the rest of our time fighting it. We spend our lives watching the weather. I wake up in the morning and pray for rain because my crops need water to grow. We are attuned to patterns in the weather that we can take advantage of. These practices started with things like the *Farmer's Almanac,* and today they include things like the algorithm we developed to help farmers know the best time to seed their crops. We take the long view when it comes to weather and climate, and we're humble enough to know we can't control these forces.

Our partnership with nature is also why so many of us in commercial agriculture have decided to adopt zero tillage and direct seeding practices, which sequester carbon in the soil. Many organic farmers still till the soil, disrupting the very organic matter we are all meant to protect. The land is our livelihood. We want to ensure that it continues to be a healthy, sustainable resource for future generations.

Today, I'm working the same land that my ancestors worked—my father, my grandfather, and my great-grandfather—and I can feel their presence. I work this land with my wife and my kids, who will take over after I retire. Therefore, for me, there is a spiritual aspect to farming.

I derive enormous satisfaction from knowing that I'm feeding people, and even though I don't see where my wheat ends up, I know that eventually, it will be on somebody's table as bread. It's going to help fill someone's belly. Maybe it's a family sitting down together on a Sunday night or a couple on a first date at a fancy restaurant in Toronto. Maybe it's someone on the other side of the planet who doesn't know how far their food has journeyed.

While it does hurt me when people accuse me of wrongdoing simply for being a commercial farmer using GMOS and glyphosate or growing canola, I want people to understand that I am grateful for the privilege of being a farmer. I consider myself lucky to wake up every morning and know I am making a difference in the world by feeding people. There is something immensely gratifying in that awareness—something that you can't get from money.

I live in a beautiful place. I'm surrounded by nature, and I work with nature every day. I am exposed to the beauty and the cruelty of nature and the seasons. All of us who are exposed to nature in this way, the people who dedicate our lives to farming, are more resilient because of it. And we have incredible respect for the environment. Off-the-cuff remarks about farmers hurting the environment and poisoning the land and wildlife certainly make me raise my eyebrows. There is abundant, diverse wildlife living on my land and even raising their young on my farm, and I take great joy in observing animals in their natural habitat.

One of my favorites is the burrowing owl. It is a small sandy-colored owl with long legs and bright yellow eyes. It lives underground in tunnels and burrows abandoned by ground squirrels, badgers, and prairie dogs. Another unique feature of the burrowing owl is the rattlesnake-like sound they make when someone approaches their burrow. You can bet I jumped back the first time I heard that rattle on the farm!

This owl is listed as an endangered species in Canada and certain US states, and it is thought the population decline is a result of habitat loss and human activities. Though the primary reason for the decline according to the U.S. Fish and Wildlife Service is deliberate pest control programs targeted at eliminating prairie dogs (who help create the burrows the owls later live in).[198] It is estimated that only 1,000 pairs exist in Canada today.

I didn't see any burrowing owls back when my dad tilled the soil, which disrupted their habitat. However, I have seen them often since I started minimal tillage on the farm. As I was writing this book, I was also harvesting fall rye, and guess what? A burrowing owl flew right in front of me as I passed by as if to say, "Thank you for providing me with a safe place to raise my family!" Maybe I was lucky, or maybe it is a sign the species is rejuvenating. Either way, you can't imagine the satisfaction of seeing beautiful, wild creatures on your land. It provided me with the gratification that we, as farmers, are progressing in the right direction.

The biodiversity doesn't stop there! In the last year, I have seen snow geese, eagles, hummingbirds, finches, red-winged blackbirds, ravens, crows, magpies, hawks, prairie chickens, sandhill cranes, mallard ducks, coots, mud hens, and many more. There are also white-tailed deer, mule deer, moose, elk, and even antelope calmly foraging on my grasslands and crops. I have seen black bears, wolves, coyotes, bobcats, and red foxes. There are Richardson ground squirrels, moles, skunks, porcupines, rabbits, jackrabbits, beavers, muskrats, weasels, and mink all around me. Bumblebees, wasps, hornets, beetles, and numerous other beneficial insects. There are earthworms, mice, garter snakes, and bats. It's an extraordinary menagerie of life.

I have even tested for the DNA of soil microorganisms, and there are too many of these present to mention! My DNA soil health summary says the total microbial activity in the soil ranges from 11,000 to 16,000, depending on soil type! What does the number mean? Well, I can tell you this: it means I have very biologically active soil, well above the minimum standard for healthy soil.

I am proud and happy to share my land with so much wildlife.

---

198 "Burrowing Owl." *U.S. Fish and Wildlife Service.* https://www.fws.gov/species/burrowing-owl-athene-cunicularia

I'd like to share one more story about the land before signing off. If you look at a very old map of Saskatchewan, you'll see a dot that marks Lizard Lake. The lake is just a couple of miles long, and the dot on the map might make you think there used to be a town there. But there was never any town. The dot marked the post office, which was really just our farmhouse. My dad served as the Lizard Lake postmaster. That meant that we *were* the post office when I was growing up.

Every Friday, my dad would drive to the town of Biggar, collect the mail, and bring it back to our house. Our porch was lined with homemade wooden mailboxes, each with a different neighbor's name carved into it, and my dad would "deliver" the mail accordingly. The neighbors, who were all farmers, would come by in the evening to get their mail, have a glass of whiskey with my father, and exchange the news of the week. That is the world I grew up in, and it was a privilege.

Plenty has changed, but plenty hasn't. I can still stand on the hill where my father's ashes are buried and look out at the beautiful countryside. The air is clear, the stars at night are as bright as you'll ever see them, and in winter, the Northern Lights dance across the horizon. It isn't a flashy lifestyle, but I think we're doing pretty good.

Our commitment as farmers to feeding people, creating jobs, contributing to the economy, protecting the land, and being reliable guardians of the food supply—that hasn't changed a lick.

Chapter 14

FINAL THOUGHTS: WE'RE LIVING LONGER AND LIVING HEALTHIER ... THANK A FARMER!

I hope this book has opened your eyes to the ways the media, influencers, politicians, and others have created an erroneous impression about the safety of the world's food supply. We've never enjoyed so much variety and safety in that supply, and we've never experienced such awesome longevity as we do in today's world. And it's not because people are eating organic or because we banned GMOs! The food supply system is safe because farmers are human beings who want the best for the people they serve—their families and yours.

We all want to live safe, healthy, and fulfilling lives, and we want the same thing for our loved ones. It can be difficult to make sense of the millions of articles about health and food safety that are online and in the news. Every other day there is a new superfood, a new diet craze, or a new toxic food you must avoid. To add to the confusion, many nutrition claims are contradictory—one day a cup of coffee is good for your heart, and the next day it will kill you. One influencer might suggest a banana and peanut butter as a healthy afternoon snack, while the next one says you should only ever eat a quarter of a banana because bananas are full of sugar and carbohydrates.

Stories and posts about our health get high engagement—which translates to ad revenue and partnership—because media and influencers know how much we value and safeguard our health. The more fear they

139

can instill in us, the more willing we are to pay whatever it takes if we think it will insulate us from the nebulous threat being peddled, or if it promises us a "beach body" for summer or some other fantasy.

People who go down this road can end up paying hundreds or even thousands of dollars for vitamins and supplements with dubious health claims or buying organic fruits and vegetables at four to five times the price of non-organic food. But there is no silver bullet, and there are no miracle cures for every ailment under the sun.

The real solution is staring us all in the face: eating an affordable, balanced diet of ordinary, commercially produced foods, getting a little exercise and a good night's sleep on a regular basis, and limiting alcohol intake is the best recipe for a healthy lifestyle.

We've covered a lot of ground in this book, and I hope it has helped you separate the wheat from the chaff and the myths from the facts. We talked about how GMO foods are not only safe but also part of the natural evolution of plant breeding, and they represent a huge leap forward for sustainable farming, especially when it comes to protecting the all-important topsoil. GMO foods like Golden Rice also have the potential to end life-threatening diseases, malnutrition, and starvation in the poorest regions of the world.

We know that "organic" doesn't mean "healthier," and that there are actually significant drawbacks to organic growing, like lower yields and higher costs, which make organic growing unrealistic on a global scale. We've also learned that organic growers still use fertilizers and pesticides, which aren't necessarily subject to the same level of scrutiny and continuous testing as synthetic inputs.

Our discussion examined what it takes to produce enough food to feed eight billion people and counting. That means employing fertilizers, pesticides, and other inputs using the 4R method—the right input, at the right rate, at the right time, and in the right place. We demystified the next stage in food production and clarified how important preservatives and chemical additives are to protecting the food supply. We also explored the extensive regulatory and enforcement bodies that ensure every part of every process in food production meets the most up-to-date safety standards.

And we know that we are living longer, healthier lives than ever before. Rates of all types of cancer have been steadily declining for decades, and average life expectancy is steadily rising. To put a finer point on it, the average life expectancy when I was born in 1956 was 68. Well, I'm 68 now, and I'm very relieved to know that the average life expectancy today is 79. Positive trends in health and longevity are emblematic of the safety and security of our food supply.

I also shared a lot about myself and what I spend my days thinking about—mostly how to keep improving my methods and protecting the land. As a fourth-generation farmer, I care deeply for the land and the environment. It's the lifeblood of what I do, and I am proud to say that I am leaving my farm in better condition than I received it in. I have also used technology to monitor and improve the health of my soil and crops.

I sometimes joke that we feed our crops better than we feed ourselves. We balance the diet of our plants to a tee. We test the soil and the plant tissue. We measure out the nutrients down to the ounce. And we adjust inputs based on our results to create the highest quality, most nutritious food we can. We do this for every plant—wheat, barley, canola, and rye—every single year. As new technologies come along, our results will get better and better, and when my daughters eventually take over running the farm, I hope it will continue to be a thriving family business.

My concern, curiosity, and passion are shared by pretty much everyone I know who does this. Most farms in North America are family-owned like mine. We have a vested interest in preserving our natural resources for future generations, for our kids and our grandkids. We need the land we work to be healthy and productive, and we spend our lives learning the best ways to win the battle against the forces of Mother Nature—harsh weather, pests, diseases, and weeds.

I don't doubt there are bad actors or poor stewards of the land out there, just as there are bad doctors or lawyers. But there is no nefarious grand conspiracy of farmers trying to make you sick to get kickbacks from Big Pharma. For the most part, farmers are committed to our work of serving others. We want what everyone else wants: a safe world for our families and friends, enough money to put food on the table and a roof over our heads, and time to enjoy this beautiful life we have.

Now, that doesn't mean that you shouldn't ask questions. I believe you should always be curious and do your due diligence. That's why I wrote this book. I think it is important to ask these questions, and I wanted to open the line of communication between farmers and consumers. My goal is that you come away with the same confidence and trust in our food producers and our food supply that I have. I hope you go to bed at night knowing that your food supply is the safest it has ever been, regardless of where you buy it.

It is a tremendously positive thing that just two percent of people in the world are farmers because that means the other 98 percent can focus on being whatever they want to be—artists, engineers, nurses, teachers, etc.—and count on having a healthy, robust food supply. It's just us farmers, a sliver of the population, who need to worry about producing enough food to feed everyone on the planet.

In this sense, farmers are the gatekeepers. The question is, do you trust us to be sustainable, practical, morally ethical, and environmentally friendly in our choices? Because in my experience most farmers are exactly that.

I am reminded of the "Cross of Gold" speech William Jennings Bryan made at the US Democratic National Convention when he ran for president in 1896. He said, "You come to us and tell us that the great cities are in favor of the gold standard; we reply that the great cities rest upon our broad and fertile prairies. Burn down your cities and leave our farms, and your cities will spring up again as if by magic, but destroy our farms, and the grass will grow in the streets of every city in the country."[199]

What he said is just as true today as it was 128 years ago. Everyone, the world over, is counting on farms and farmers without even realizing it. Farms and farmers literally keep us alive. And it's an increasing challenge with the prospect of nine billion people to feed and nourish. Global food security is a major issue in the 21st century, and we are doing everything we can to meet that challenge.

We provide the food people need to survive, yes, but there is much more to it than that. Food is part of our celebrations and culture. It's chicken noodle soup when you're sick, and burgers and hot dogs at your summer barbeque. It's birthday cake and Thanksgiving dinner. It's a ribeye steak and mashed potatoes to celebrate your promotion at work, a hearty breakfast before the big game, or your grandma's chocolate chip cookies when you visit her.

Remember the convention host I mentioned at the outset? Well, I invited him up to my farm this past summer. We had many meaningful conversations, and I even introduced him to some heavyweights in Big Agriculture! And I like to think we both learned something from each other.

I hope that you, like that individual, will now have greater trust in the food supply system. I hope that you'll lobby your politicians to support the introduction of Golden Rice into Africa to prevent childhood starvation. I hope that you won't spend more money than you need to buy organic because it's "safe." Pretty much everything that farmers produce is safe, or it wouldn't be on the market. That's what farmers—this one, and all his colleagues—want you to know about food!

Bon appétit!

199 "William Jennings Bryan, Democratic National Convention Address." *American Rhetoric, Online Speech Bank.* Originally delivered 8 July 1896; site updated 10 December 2021. https://www.americanrhetoric.com/speeches/williamjenningsbryan1896dnc.htm

ABOUT THE AUTHOR

Dennis Bulani is a fourth-generation farmer and business owner in Saskatchewan, Canada. His debut book, *What a Farmer Wants You to Know About Food*, seeks to add the farmer's voice to the mainstream conversation about where our food comes from. From GMOs to pesticides, from preservatives to fertilizers—and even the ethics of food production, *What a Farmer Wants You to Know About Food* answers the questions you have (and probably a few you didn't know you had!).

ACKNOWLEDGEMENTS

This book reflects conversations I've been having most of my life – some of them quite regularly with friends outside of farming. As a farmer, I've been able to draw on my own experiences in these exchanges, but I didn't have ready access to the data, studies and resources that make people feel comfortable and reassured about farming and food.

I always told myself I'd pull all that information together one day… and now, I've finally done it! I hope this book serves as a tool for farmers to help answer the great questions consumers ask about food.

I've spoken candidly in these pages. I'm not trying to sell a diet, a program, or a supplement – I just want to provide frank, honest answers to questions I've been asked and to reach more of the people asking them. I do hope to sell several copies of this book, if only to have others join the conversation!

I want to acknowledge my wife, Lynda, and my three daughters, Chelsea, Cassandra and Coralee, who have grown up in the farming world and have served as witnesses and sometimes participants in these discussions – we've had conversations, debates, and maybe even a few monologues from me. They recognize agriculture is in my blood; it's a part of my heritage.

I also want to acknowledge Dr. Aron Cory, Shannon Walker, Troy LaForge, Moira Knight and Chelsea Norheim for assisting me in helping to proof, edit and research some of the topics in these pages. Their insights and constructive remarks were invaluable. And, of course, a big thank you to my friends and colleagues in the agriculture industry who helped me clarify and confirm details and concepts.

Thanks to Mary Fearon, for seeing and sharing my vision and supporting me through the journey of bringing it to life. And to her team at OnPrpose for helping me get the book out in the world.

Thanks to Michael Levin, a master of his craft, for his partnership in bringing this book to life. It is the result of his ability to find, shape and capture the story inside me.

Thank you to Patricia Bacall Garver, my dedicated and patient book designer and publishing consultant for the beautiful design and production of this book.

Finally, thank you to my farmer friends and customers for allowing me to be a small part in your lives.